Mindset Mathematics

Visualizing and Investigating Big Ideas

Jo Boaler

Jen Munson

Cathy Williams

JOSSEY-BASS™

A Wiley Brand

Published by Jossey-Bass
A Wiley Brand
111 River St, Hoboken, NJ 07030
www.josseybass.com

Jossey-Bass books and products are available through most bookstores. To contact Jossey-Bass directly call our Customer Care Department within the U.S. at 800-956-7739, outside the U.S. at 317-572-3986, or fax 317-572-4002.

Wiley publishes in a variety of print and electronic formats and by print-on-demand. Some material included with standard print versions of this book may not be included in e-books or in print-on-demand. If this book refers to media such as a CD or DVD that is not included in the version you purchased, you may download this material at http://booksupport.wiley.com. For more information about Wiley products, visit www.wiley.com.

The Visualize, Play, and Investigate icons are used under license from Shutterstock.com and the following arists: Blan-k, Marish, and SuzanaM.

Library of Congress Cataloging-in-Publication Data is Available:

ISBN 9781119358626 (PaperBack)
ISBN 9781119358756 (epdf)
ISBN 9781119358732 (epub)

Cover design by Wiley
Cover image: © Marish/Shutterstock-Eye; © Kritchanut/iStockphoto- Background

FIRST EDITION

SKY10045441_040623

Contents

To all those teachers pursuing a mathematical mindset journey with us.

Introduction

I still remember the moment when Youcubed, the Stanford center I direct, was conceived. I was at the Denver NCSM and NCTM conferences in 2013, and I had arranged to meet Cathy Williams, the director of mathematics for Vista Unified School District. Cathy and I had been working together for the past year improving mathematics teaching in her district. We had witnessed amazing changes taking place, and a filmmaker had documented some of the work. I had recently released my online teacher course, called How to Learn Math, and been overwhelmed by requests from tens of thousands of teachers to provide them with more of the same ideas. Cathy and I decided to create a website and use it to continue sharing the ideas we had used in her district and that I had shared in my online class. Soon after we started sharing ideas on the Youcubed website, we were invited to become a Stanford University center, and Cathy became the codirector of the center with me.

In the months that followed, with the help of one of my undergraduates, Montse Cordero, our first version of youcubed.org was launched. By January 2015, we had managed to raise some money and hire engineers, and we launched a revised version of the site that is close to the site you may know today. We were very excited that in the first month of that relaunch, we had five thousand visits to the site. At the time of writing this, we are now getting three million visits to the site each month. Teachers are excited to learn about the new research and to take the tools, videos, and activities that translate research ideas into practice and use them in their teaching.

Low-Floor, High-Ceiling Tasks

One of the most popular articles on our website is called "Fluency without Fear." I wrote this with Cathy when I heard from many teachers that they were being made to use timed tests in the elementary grades. At the same time, new brain science was emerging showing that when people feel stressed—as students do when facing a timed test—part of their brain, the working memory, is restricted. The working memory is exactly the area of the brain that comes into play when students need to calculate with math facts, and this is the exact area that is impeded when students are stressed. We have evidence now that suggests strongly that timed math tests in the early grades are responsible for the early onset of math anxiety for many students. I teach an undergraduate class at Stanford, and many of the undergraduates are math traumatized. When I ask them what happened to cause this, almost all of them will recall, with startling clarity, the time in elementary school when they were given timed tests. We are really pleased that "Fluency without Fear" has now been used across the United States to pull timed tests out of school districts. It has been downloaded many thousands of times and used in state and national hearings.

One of the reasons for the amazing success of the paper is that it does not just share the brain science on the damage of timed tests but also offers an alternative to timed tests: activities that teach math facts conceptually and through activities that students and teachers enjoy. One of the activities—a game called How Close to 100—became so popular that thousands of teachers tweeted photos of their students playing the game. There was so much attention on Twitter and other media that Stanford noticed and decided to write a news story on the damage of speed to mathematics learning. This was picked up by news outlets across the United States, including *US News & World Report,* which is part of the reason the white paper has now had so many downloads and so much impact. Teachers themselves caused this mini revolution by spreading news of the activities and research.

How Close to 100 is just one of many tasks we have on youcubed.org that are extremely popular with teachers and students. All our tasks have the feature of being "low floor and high ceiling," which I consider to be an extremely important quality for engaging all students in a class. If you are teaching only one student, then a mathematics task can be fairly narrow in terms of its content and difficulty. But whenever you have a group of students, there will be differences in their needs, and they will be challenged by different ideas. A low-floor, high-ceiling task is one in which everyone can engage, no matter what his or her prior understanding or knowledge, but also one that is open enough to extend to high levels, so that all students can be deeply challenged. In the last two years, we have launched an introductory week of mathematics lessons on our site that are open, visual, and low floor, high ceiling. These have been extremely popular with teachers; they have had approximately four million downloads and are used in 20% of schools across the United States.

In our extensive work with teachers around the United States, we are continually asked for more tasks that are like those on our website. Most textbook publishers seem to ignore or be unaware of research on mathematics learning, and most textbook questions are narrow and insufficiently engaging for students. It is imperative that the new knowledge of the ways our brains learn mathematics is incorporated into the lessons students are given in classrooms. It is for this reason that we chose to write a series of books that are organized around a principle of active student engagement, that reflect the latest brain science on learning, and that include activities that are low floor and high ceiling.

Youcubed Summer Camp

We recently brought 81 students onto the Stanford campus for a Youcubed summer math camp, to teach them in the ways that are encouraged in this book. We used open, creative, and visual math tasks. After only 18 lessons with us, the students improved their test score performance by an average of 50%, the equivalent of 1.6 years of school. More important, they changed their relationship with mathematics and started believing in their own potential. They did this, in part, because we talked to them about the brain science showing that

- There is no such thing as a math person—anyone can learn mathematics to high levels.
- Mistakes, struggle, and challenge are critical for brain growth.
- Speed is unimportant in mathematics.
- Mathematics is a visual and beautiful subject, and our brains want to think visually about mathematics.

All of these messages were key to the students' changed mathematics relationship, but just as critical were the tasks we worked on in class. The tasks and the messages about the brain were perfect complements to each other, as we told students they could learn anything, and we showed them a mathematics that was open, creative, and engaging. This approach helped them see that they could learn mathematics and actually do so. This book shares the kinds of tasks that we used in our summer camp, that make up our week of inspirational mathematics (WIM) lessons, and that we post on our site.

Before I outline and introduce the different sections of the book and the ways we are choosing to engage students, I will share some important ideas about how students learn mathematics.

Memorization versus Conceptual Engagement

Many students get the wrong idea about mathematics—exactly the wrong idea. Through years of mathematics classes, many students come to believe that their role in mathematics learning is to memorize methods and facts, and that mathematics success comes from memorization. I say this is exactly the wrong idea because there is actually very little to remember in mathematics. The subject is made up of a few big, linked ideas, and students who are successful in mathematics are those who see the subject as a set of ideas that they need to think deeply about. The Program for International Student Assessment (PISA) tests are international assessments of mathematics, reading, and science that are given every three years. In 2012, PISA not only assessed mathematics achievement but also collected data on students' approach to mathematics. I worked with the PISA team in Paris at the Organisation for Economic Co-operation and Development (OECD) to analyze students' mathematics approaches and their relationship to achievement. One clear result emerged from this analysis. Students approached mathematics in three distinct ways. One group approached mathematics by attempting to memorize the methods they had met; another group took a "relational" approach, relating new concepts to those they already knew; and a third group took a self-monitoring approach, thinking about what they knew and needed to know.

In every country, the memorizers were the lowest-achieving students, and countries with high numbers of memorizers were all lower achieving. In no country were memorizers in the highest-achieving group, and in some high-achieving countries such as Japan, students who combined self-monitoring and relational strategies outscored memorizing students by more than a year's worth of schooling. More detail on this finding is given in this *Scientific American* Mind article that I coauthored with a PISA analyst: https://www.scientificamerican.com/article/ why-math-education-in-the-u-s-doesn-t-add-up/.

Mathematics is a conceptual subject, and it is important for students to be thinking slowly, deeply, and conceptually about mathematical ideas, not racing through methods that they try to memorize. One reason that students need to think conceptually has to do with the ways the brain processes mathematics. When we learn new mathematical ideas, they take up a large space in our brain as the brain works out where they fit and what they connect with. But with time, as we move on with our understanding, the knowledge becomes compressed in the brain, taking up a very small space. For first graders, the idea of addition takes up a large space in their brains as they think about how it works and what it means, but for adults the idea of addition is compressed, and it takes up a small space. When adults are asked to add 2 and 3, for example, they can quickly and easily extract the compressed knowledge. William Thurston (1990), a mathematician who won the Field's Medal—the highest honor in mathematics—explains compression like this:

> Mathematics is amazingly compressible: you may struggle a long time, step by step, to work through the same process or idea from several approaches. But once you really understand it and have the mental perspective to see it as a whole, there is often a tremendous mental compression. You can file it away, recall it quickly and completely when you need it, and use it as just one step in some other mental process. The insight that goes with this compression is one of the real joys of mathematics.

You will probably agree with me that not many students think of mathematics as a "real joy," and part of the reason is that they are not compressing mathematical ideas in their brain. This is because the brain only compresses concepts, not methods. So if students are thinking that mathematics is a set of methods to memorize, they are on the wrong pathway, and it is critical that we change that. It is very important that students think deeply and conceptually about ideas. We provide the activities in this book that will allow students to think deeply and conceptually, and an essential role of the teacher is to give the students time to do so.

Mathematical Thinking, Reasoning, and Convincing

When we worked with our Youcubed camp students, we gave each of them journals to record their mathematical thinking. I am a big fan of journaling—for myself and my students. For mathematics students, it helps show them that mathematics is a subject for which we should record ideas and pictures. We can use journaling to encourage students to keep organized records, which is another important part of mathematics, and help them understand that mathematical thinking can be a long and slow process. Journals also give students free space—where they can be creative, share ideas, and feel ownership of their work. We did not write in the students' journals, as we wanted them to think of the journals as their space, not something that teachers wrote on. We gave students feedback on sticky notes that we stuck onto their work. The images in Figure I.1 show some of the mathematical records the camp students kept in their journals.

Another resource I always share with learners is the act of color-coding—that is, students using colors to highlight different ideas. For example, when working on an algebraic task, they may show the x in the same color in an expression, in a graph, and in a picture, as shown in Figure I.2. When adding numbers, color-coding may help show the addends (Figure I.3).

Color-coding highlights connections, which are a really critical part of mathematics.

Another important part of mathematics is the act of reasoning—explaining why methods are chosen and how steps are linked, and using logic to connect ideas.

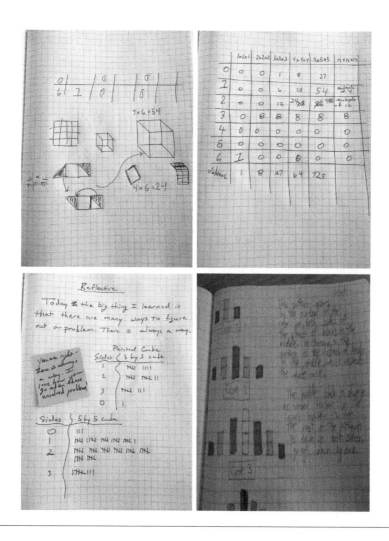

Figure I.1

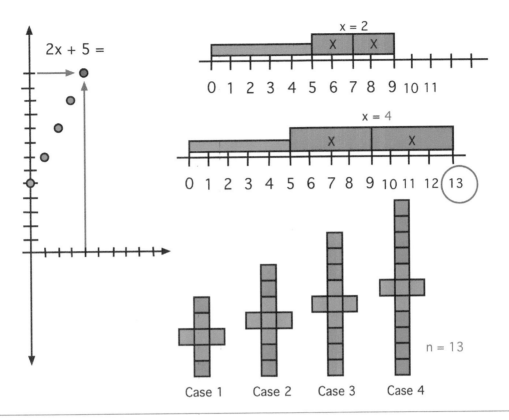

Figure I.2

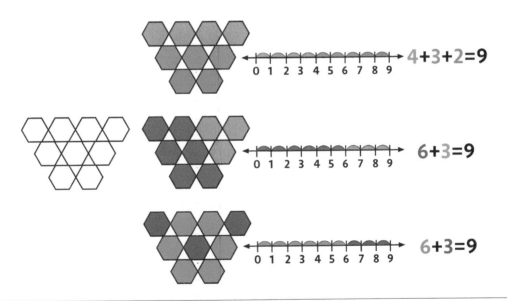

Figure I.3

Reasoning is at the heart of mathematics. Scientists prove ideas by finding more cases that fit a theory, or countercases that contradict a theory, but mathematicians prove their work by reasoning. If students are not reasoning, then they are not really doing mathematics. In the activities of these books, we suggest a framework that encourages students to be convincing when they reason. We tell them that there are three levels of being convincing. The first, or easiest, level is to convince yourself of something. A higher level is to convince a friend. And the highest level of all is to convince a skeptic. We also share with students that they should be skeptics with one another, asking one another why methods were chosen and how they work. We have found this framework to be very powerful with students; they enjoy being skeptics, pushing each other to deeper levels of reasoning, and it encourages students to reason clearly, which is important for their learning.

We start each book in our series with an activity that invites students to reason about mathematics and be convincing. I first met an activity like this when reading Mark Driscoll's teaching ideas in his book *Fostering Algebraic Thinking*. I thought it was a perfect activity for introducing the skeptics framework that I had learned from a wonderful teacher, Cathy Humphreys. She had learned about and adapted the framework from two of my inspirational teachers from England: mathematician John Mason and mathematics educator Leone Burton. As well as encouraging students to be convincing, in a number of activities we ask students to prove an idea. Some people think of proof as a formal set of steps that they learned in geometry class. But the act of proving is really about connecting ideas, and as students enter the learning journey of proving, it is worthwhile celebrating their steps toward formal proof. Mathematician Paul Lockhart (2012) rejects the idea that proving is about following a set of formal steps, instead proposing that proving is "abstract art, pure and simple. And art is always a struggle. There is no systematic way of creating beautiful and meaningful paintings or sculptures, and there is also no method for producing beautiful and meaningful mathematical arguments" (p. 8). Instead of suggesting that students follow formal steps, we invite them to think deeply about mathematical concepts and make connections. Students will be given many ways to be creative when they prove and justify, and for reasons I discuss later, we always encourage and celebrate visual as well as numerical and algebraic justifications. Ideally, students will create visual, numerical, and algebraic representations and connect their ideas through color-coding and through verbal explanations. Students are excited to experience mathematics in these ways, and they benefit from the opportunity to bring their individual ideas and creativity to the problem-solving and learning space. As students develop in their mathematical understanding, we can encourage them to extend and generalize their ideas through reasoning, justifying, and proving. This process deepens their understanding and helps them compress their learning.

Big Ideas

The books in the Mindset Mathematics Series are all organized around mathematical "big ideas." Mathematics is not a set of methods; it is a set of connected ideas that need to be understood. When students understand the big ideas in mathematics, the methods and rules fall into place. One of the reasons any set of curriculum standards is flawed is that standards take the beautiful subject of mathematics and its many connections, and divide it into small pieces that make the connections disappear. Instead of starting with the small pieces, we have started with the big ideas and important connections, and have listed the relevant Common Core curriculum standards within the activities. Our activities invite students to engage in the mathematical acts that are listed in the imperative Common Core practice standards, and they also teach many of the Common Core content standards, which emerge from the rich activities. Student activity pages are noted with a ⊕ and teacher activity pages are noted with a ⊕.

Although we have chapters for each big idea, as though they are separate from each other, they are all intrinsically linked. Figure I.4 shows some of the connections between the ideas, and you may be able to see others. It is very important to share with students that mathematics is a subject of connections and to highlight the connections as students work. You may want to print the color visual of the different connections for students to see as they work. To see the maps of big ideas for all of the grades K through 8, find our paper "What Is Mathematical Beauty?" at youcubed.org.

Structure of the Book

Visualize. Play. Investigate. These three words provide the structure for each book in the series. They also pave the way for open student thinking, for powerful brain connections, for engagement, and for deep understanding. How do they do that? And why is this book so different from other mathematics curriculum books?

Visualize 🌀

For the past few years, I have been working with a neuroscience group at Stanford, under the direction of Vinod Menon, which specializes in mathematics learning. We have been working together to think about the ways that findings from brain science can be used to help learners of mathematics. One of the exciting discoveries that has

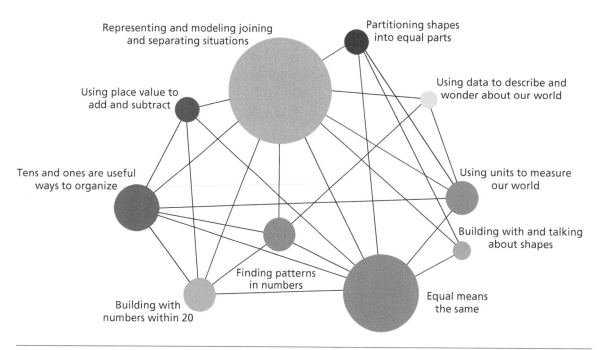

Figure I.4

been emerging over the last few years is the importance of visualizing for the brain and our learning of mathematics. Brain scientists now know that when we work on mathematics, even when we perform a bare number calculation, five areas of the brain are involved, as shown in Figure I.5.

Two of the five brain pathways—the dorsal and ventral pathways—are visual. The dorsal visual pathway is the main brain region for representing quantity. This may seem surprising, as so many of us have sat through hundreds of hours of mathematics classes working with numbers, while barely ever engaging visually with mathematics. Now brain scientists know that our brains "see" fingers when we calculate, and knowing fingers well—what they call finger perception—is critical for the development of an understanding of number. If you would like to read more about the importance of finger work in mathematics, look at the visual mathematics section of youcubed.org. Number lines are really helpful, as they provide the brain with a visual representation of number order. In one study, a mere four 15-minute sessions of students playing with a number line completely eradicated the differences between students from low-income and middle-income backgrounds coming into school (Siegler & Ramani, 2008).

Our brain wants to think visually about mathematics, yet few curriculum materials engage students in visual thinking. Some mathematics books show pictures, but they rarely ever invite students to do their own visualizing and drawing. The neuroscientists' research shows the importance not only of visual

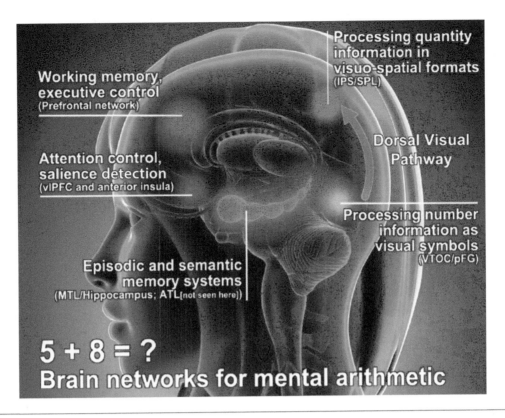

Figure I.5

thinking but also of students' connecting different areas of their brains as they work on mathematics. The scientists now know that as children learn and develop, they increase the connections between different parts of the brain, and they particularly develop connections between symbolic and visual representations of numbers. Increased mathematics achievement comes about when students are developing those connections. For so long, our emphasis in mathematics education has been on symbolic representations of numbers, with students developing one area of the brain that is concerned with symbolic number representation. A more productive and engaging approach is to develop all areas of the brain that are involved in mathematical thinking, and visual connections are critical to this development.

In addition to the brain development that occurs when students think visually, we have found that visual activities are really engaging for students. Even students who think they are "not visual learners" (an incorrect idea) become fascinated and think deeply about mathematics that is shown visually—such as the visual representations of the calculation 18 × 5 shown in Figure I.6.

In our Youcubed teaching of summer school to sixth- and seventh-grade students and in our trialing of Youcubed's WIM materials, we have found that students are inspired by the creativity that is possible when mathematics is

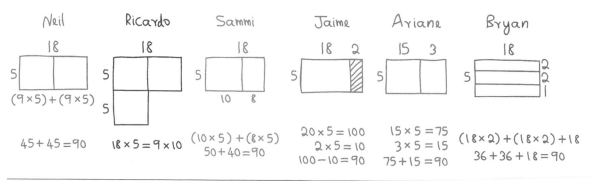

Figure I.6

visual. When we were trialing the materials in a local middle school one day, a parent stopped me and asked what we had been doing. She said that her daughter had always said she hated and couldn't do math, but after working on our tasks, she came home saying she could see a future for herself in mathematics. We had been working on the number visuals that we use throughout these teaching materials, shown in Figure I.7.

The parent reported that when her daughter had seen the creativity possible in mathematics, everything had changed for her. I strongly believe that we can give these insights and inspirations to many more learners with the sort of creative, open mathematics tasks that fill this book.

We have also found that when we present visual activities to students, the status differences that often get in the way of good mathematics teaching disappear. I was visiting a first-grade classroom recently, and the teacher had set up four different stations around the room. In all of them, the students were working on arithmetic. In one, the teacher engaged students in a mini number talk; in another, a teaching assistant worked on an activity with coins; in the third, the students played a board game; and in the fourth, they worked on a number worksheet. In each of the first three stations, the students collaborated and worked really well, but as soon as students went to the worksheet station, conversations changed, and in every group I heard statements like "This is easy," "I've finished," "I can't do this," and "Haven't you finished yet?" These status comments are unfortunate and off-putting for many students. I now try to present mathematical tasks without numbers as often as possible, or I take out the calculation part of a task, as it is the numerical and calculational aspects that often cause students to feel less sure of themselves. This doesn't mean that students cannot have a wonderful and productive relationship with numbers, as we hope to promote in this book, but sometimes the key mathematical idea can be arrived at without any numbers at all.

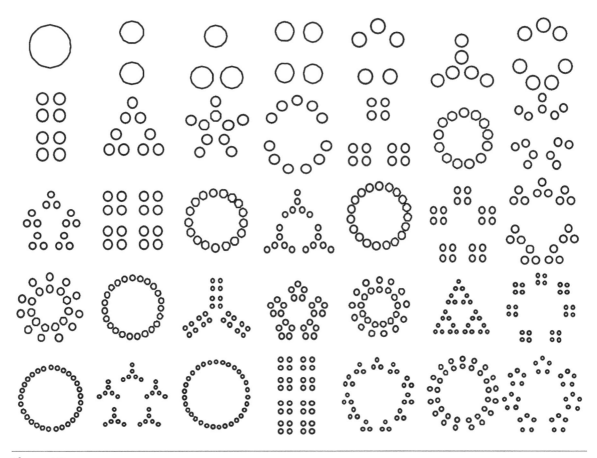

Figure I.7

Almost all the tasks in our book invite students to think visually about mathematics and to connect visual and numerical representations. This encourages important brain connections as well as deep student engagement.

Play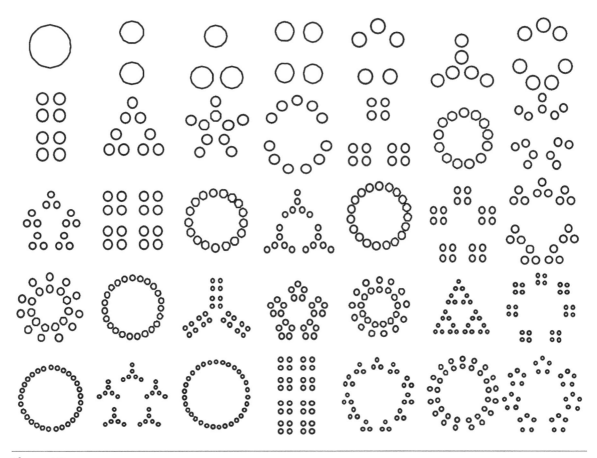

The key to reducing status differences in mathematics classrooms, in my view, comes from *opening* mathematics. When we teach students that we can see or approach any mathematical idea in different ways, they start to respect the different thinking of all students. Opening mathematics involves inviting students to see ideas differently, explore with ideas, and ask their own questions. Students can gain access to the same mathematical ideas and methods through creativity and exploration that they can by being taught methods that they practice. As well as reducing or removing status differences, open mathematics is more engaging for students. This is why we are inviting students, through these mathematics materials, to play with mathematics.

Albert Einstein famously once said that "play is the highest form of research." This is because play is an opportunity for ideas to be used and developed in the service of something enjoyable. In the Play activities of our materials, students are invited to work with an important idea in a free space where they can enjoy the freedom of mathematical play. This does not mean that the activities do not teach essential mathematical content and practices—they do, as they invite students to work with the ideas. We have designed the Play activities to downplay competition and instead invite students to work with each other, building understanding together.

Investigate ❓

Our Investigate activities add something very important: they give students opportunities to take ideas to the sky. They also have a playful element, but the difference is that they pose questions that students can explore and take to very high levels. As I mentioned earlier, all of our tasks are designed to be as low floor and high ceiling as possible, as these provide the best conditions for engaging all students, whatever their prior knowledge. Any student can access them, and students can take the ideas to high levels. We should always be open to being surprised by what our learners can do, and always provide all students with opportunities to take work to high levels and to be challenged.

A crucial finding from neuroscience is the importance of students struggling and making mistakes—these are the times when brains grow the most. In one of my meetings with a leading neuroscientist, he stated it very clearly: if students are not struggling, they are not learning. We want to put students into situations where they feel that work is hard, but within their reach. Do not worry if students ask questions that you don't know the answer to; that is a good thing. One of the damaging ideas that teachers and students share in education is that teachers of mathematics know everything. This gives students the idea that mathematics people are those who know a lot and never make mistakes, which is an incorrect and harmful message. It is good to say to your students, "That is a great question that we can all think about" or "I have never thought about that idea; let's investigate it together." It is even good to make mistakes in front of students, as it shows them that mistakes are an important part of mathematical work. As they investigate, they should be going to places you have never thought about—taking ideas in new directions and exploring uncharted territory. Model for students what it means to be a curious mathematics learner, always open to learning new ideas and being challenged yourself.

* * *

We have designed activities to take at least a class period, but some of them could go longer, especially if students ask deep questions or start an investigation into a cool idea. If you can be flexible about students' time on activities, that is ideal, or you may wish to suggest that students continue activities at home. In our teaching of these activities, we have found that students are so excited by the ideas that they take them home to their families and continue working on them, which is wonderful. At all times, celebrate deep thinking over speed, as that is the nature of real mathematical thought. Ask students to come up with creative representations of their ideas; celebrate their drawing, modeling, and any form of creativity. Invite your students into a journey of mathematical curiosity and take that journey with them, walking by their side as they experience the wonder of open, mindset mathematics.

References

Lockhart, P. (2012). *Measurement*. Cambridge, MA: Harvard University Press.

Siegler, R. S., & Ramani, G. B. (2008). Playing linear numerical board games promotes low income children's numerical development. *Developmental Science, 11*(5), 655–661. doi:10.1111/j.1467-7687.2008.00714.x

Thurston, W. (1990). Mathematical education. *Notices of the American Mathematical Society, 37*(7), 844–850.

Note on Materials

Primary classrooms often have a wealth of mathematics manipulatives and materials for modeling and exploring the world. We believe, and extensive research supports, that all math learners benefit from mathematics that is visual, concrete, and modeled in multiple representations. Students need to physically create, draw, and construct mathematics to build deep understanding of what concepts represent and mean. Students need to interact with mathematics, manipulating representations to pose and investigate questions. Apps and digital games are another choice, and we have found them to be valuable because they can be organized and manipulated with an unending supply. However, we want to emphasize that they should not be a replacement for the tactile experience of working with physical manipulatives. We support different tools being available for students to use for representation as they see fit, and afterward we encourage you to ask students to reflect on what the tools allowed them to see mathematically.

Throughout our books, you will find an emphasis on visual mathematics and using manipulatives. The list that follows includes the materials that we use in the lessons in this book, along with how we see these tools as being relevant to mathematics learning. If manipulatives or materials are not available in your building, understanding the purpose of these tools may help you locate substitutes that will support students in engaging with the big ideas in this book.

Manipulatives and Materials Used in This Book

- **Snap cubes.** Snap or linking cubes are perhaps the most flexible mathematical manipulative, and we recommend these for all grade levels. In first grade, cubes can be used for counting, organizing, and modeling situations. Linking cubes together, students can see the linear quality of number, and numbers can be composed and decomposed into parts, supporting joining and separating concepts. They are particularly useful joined into groups of 10s and 1s to support ideas around place value.

- **Square tiles or chips.** Square tiles are a flexible manipulative that can be used to represent square units and build patterns from squares physically. Chips are typically circular counters, sometimes with different colors on each side. In first grade, we use these tools, along with cubes, to model joining and separating situations.

- **Rekenreks.** Rekenreks are bead frames with 10 beads in each row, typically with five beads in one color and five in another to support subitizing. Some rekenreks have just two rows, or 20 beads total, and these can be particularly useful for first graders working with joining and separating within 20. As students' thinking about joining and separating grows, they will benefit from 10-row rekenreks, which have 100 beads and support students in using place value.

- **Base 10 blocks.** Base 10 blocks are structures to represent groups of units, tens, hundreds, and thousands, such that each of these groups can be composed to represent larger groups. Base 10 blocks are typically used to model operations with larger numbers and encourage the use of place value in the development of strategies. In first grade, students may become ready to use the units and sticks of 10 after much work with building sticks of 10 themselves with snap cubes.

- **Dice.** Dice are used for game play and to explore counting and joining. Dice are wonderful tools for supporting subitizing (recognizing quantities without counting), which students will learn to do more readily with many opportunities to work with dice.

- **Tangrams.** Tangrams are sets of seven polygons that together can form a square. In first grade, students use tangrams to explore ideas about composing and decomposing shapes, as well as concepts related to equivalence and fractions.

- **Coins.** Money is a particular context for counting, skip-counting, organizing, using place value, joining, and separating. Because each coin is physically different and each represents a different number of cents, problems involving money are often best modeled using coins themselves. We recommend having plastic (or real) coins on hand for investigation in first grade.

- **Cuisenaire rods.** Cuisenaire rods are a set of rods in each of 10 lengths, from 1 unit to 10 units. The rods are unmarked, allowing students to explore the relationships between the lengths of different rods and to count their lengths in different ways. In first grade, we use Cuisenaire rods for patterning and as units for measuring lengths.

- **Bingo chips.** Bingo chips, or tinted transparent circular markers, can help students temporarily mark spaces on a page while still being able to see what each chip is covering. In first grade, we use these to support students in constructing and solving puzzles having to do with equality.

- **Colors.** Drawing is one entry point for modeling mathematical situations and for recording thinking. Later, color-coding work becomes a powerful tool to support decomposition, patterning, and connecting representations. We often ask that students have access to colors; whether they are markers, colored pencils, or colored pens, we leave up to you.

- **Collections of small objects.** Collections of objects offer students of all ages the opportunity to count, organize, sort, and estimate. In most cases, there are many different types of objects that can support this kind of mathematical work, such as beads, coins, bears, pencils, or buttons, in addition to the math manipulatives discussed earlier.

- **Tools for organizing, such as bowls or cups.** As students are developing ideas about how to sort and organize objects for counting or to understand properties, they will benefit from tools that help them maintain organization. Paper cups or bowls work well, as will other small containers or bins. We also like muffin tins for very small objects because they are more difficult to knock over.

- **Recycled supplies for building, such as cardboard boxes.** There is just not enough building in school. Children need lots of opportunities to construct, and not all students have these opportunities at home. For first grade, we recommend collecting empty boxes and containers from around the school, where deliveries are common, and from students, who can contribute empty cereal or tissue boxes. These can be used to construct larger forms and can be compared in discussions of attributes.

- **Plastic sheet protectors.** When students create their own puzzles or mathematical tasks for others to solve, placing student work inside a plastic sheet protector can allow others to mark on it with dry-erase markers without changing the original. Using sheet protectors can then enable students to exchange their creations multiple times without the need for photocopying.

- **School supplies, such as construction paper, glue sticks, sticky notes, index cards, pipe cleaners, erasers, paper clips, rulers, file folders, and masking tape.** We use these across the book to construct charts, display thinking, or piece together work. In first grade, ordinary office supplies can also be used as collections for counting and organizing.

- **Plastic gloves.** In first grade, we engage students in sorting data physically, and in one investigation, the data students explore is their own recycling or trash. Students will need gloves and any other tools (such as plastic tablecloths or plastic bins) for keeping a sanitary space as they sort through what gets recycled or thrown out in your classroom.

Activities for Building Norms

Encouraging Good Group Work

We always use this activity before students work on math together, as it helps improve group interactions. Teachers who have tried this activity have been pleased by students' thoughtful responses and have found students' thoughts and words helpful in creating a positive and supportive environment. The first thing to do is to ask students, in groups, to reflect on things they don't like people to say or do in a group when they are working on math together. Students come up with quite a few important ideas, such as not liking people to give away the answer, to rush through the work, or to ignore other people's ideas. When students have had enough time in groups brainstorming, collect the ideas. We usually do this by making a What We Don't Like list or chart and asking each group to contribute one idea, moving around the room until a few good ideas have been shared. Then we do the same for What We Do Like, creating a list or chart as a class. It can be useful to present the final charts to the class as agreed-on classroom norms that you and they can reflect back on, and add to, over the year. If any student shares a comment that casts others in a negative light, such as "I don't like waiting for slow people," do not put it on the chart; instead use it as a chance to discuss the issue and remind students that everyone's needs and ideas are important. This rarely happens, and students are usually very thoughtful and respectful in the ideas they share.

Activity	Time	Description/Prompt	Materials
Launch	5 min	Explain to students that working in groups is an important part of what mathematicians do. Mathematicians discuss their ideas and work together to solve challenging problems. It's important to work together, and we need to discuss what helps us work well together.	
Explore	10 min	In small groups, invite students to . . . 1. Reflect on the things you do not like people to say or do when you are working on math together in a group 2. Reflect on the things you do like people to say or do when you are working on math together in a group Depending on students' capacity for writing, groups can record, draw, or remember every group member's ideas. Then together the group decides which idea to share during the whole-class discussion.	• Paper • Pencil or pen
Discuss	10 min	Ask each group to share one idea for "What We Don't Like" and "What We Do Like" when working on math in groups. Collect students' ideas on a poster so that they are visible and you can refer to them during class.	Chart paper and markers

How Many Do You See? Learning to Reason, Convince, and Pose Questions

One of the most important topics in mathematics is reasoning. Scientists prove or disprove ideas by finding cases. Mathematicians prove their ideas by reasoning—making logical connections between ideas. This activity gives students an opportunity to learn to reason really well by convincing others who pose questions.

Before beginning the activity, explain to students that their role is to share their thinking and be convincing. Students may well be unfamiliar with what it means to *convince*, and you may want to give examples from students' everyday lives to illustrate what it means to be convinced of an idea. The easiest person to convince is yourself. A higher level of being convincing is to convince a friend, and the highest level of all is to convince a skeptic, someone who is unsure or doubtful. Just as important as convincing a skeptic is learning to be the skeptic. For young children, this first involves learning to attend to one another's ideas and ask questions about them. In this activity, we ask students to grapple with these three roles, making their thinking public to others, constructing arguments that convince, and listening to and questioning one another's thinking.

For this activity, we present students with an image and simply ask, "How many do you see?" On the face of it, this seems like a closed question, one with a single right answer and little reasoning. However, figuring out how many involves students in reasoning about what to count, finding ways to count when they cannot touch each object, and noticing structures in the image that can help them determine how many. The images we provide are complex, with several different types of objects that could be counted. Sometimes objects are arranged in a structured way, while other objects may be clustered randomly. Students might notice groups and use subitizing, composing, or decomposing to determine how many they think there are. Even when students arrive at the same answer for the same groups, they very well may have seen the quantities differently. We have provided four images you can use for this task, but you can find additional images in Christopher Danielson's (2018) *How Many?* or create your own out of objects in your classroom or school.

The idea in this activity is to encourage students to pay attention to and share their thinking. They must learn to listen to and wonder about the thinking of their peers. If one student is struggling to understand how another student saw the objects, encourage them to ask a question, such as, "Where did you see five?" or "Can you show me?" These are opportunities for everyone to learn that everyone's thinking is valued, that posing questions is something mathematicians do, and that explaining our thinking is how we convince one another.

Activity	Time	Description/Prompt	Materials
Launch	5 min	Tell students that they have some jobs to do as mathematicians: 1. Mathematicians reason and share their thinking with others. 2. Mathematicians explain, show, or give evidence to convince others that their thinking makes sense. 3. Mathematicians listen to the thinking of others and ask questions to make sense of their ideas. You may want to make a chart to refer back to in the future. Tell students you are going to show them an image that has lots of different things in it. Their job is to figure out how many they see.	Optional: chart and markers
Explore	10 min	Show students one of the How Many Do You See? images on the document camera. Ask, How many do you see? Give students some time to think. Invite students to signal that they are ready to share their thinking with a thumbs-up held low to the body so that others can continue to think. Tell students you are going to ask them to share their thinking and explain what they saw. Say, If anyone shares thinking that you don't yet understand, it is your job to ask a question to help them explain and to convince you. Ask, How many do you see? Invite students to share the quantities that they saw, and press students to specify what they counted or the unit attached to their number. Ask questions to support students in explaining what they saw and how they saw it, such as, "Where do you see the five?" "Five what?" and "How did you see that there were five?" Record students' thinking on the image, labeling the number, the unit, and how they saw the quantity. Regularly ask the class whether anyone has any questions, particularly if you see confused faces or hear reasoning that is difficult to follow.	How Many Do You See? image, to display

Activity	Time	Description/Prompt	Materials
Discuss	5–10 min	Refer back to the three roles that students had: reasoning, convincing, and posing questions. For each role, ask, When do you hear someone reason/convince/pose questions? Invite students to share examples from the discussion of others offering their ideas, providing evidence that was convincing, and asking questions to better understand someone else's thinking. Point out examples that you saw that students may not have recognized. Tell students that this is the kind of work that you will be asking them to do all year.	

Reference

Danielson, C. (2018). *How many?* Portland, ME: Stenhouse.

How Many Do You See?

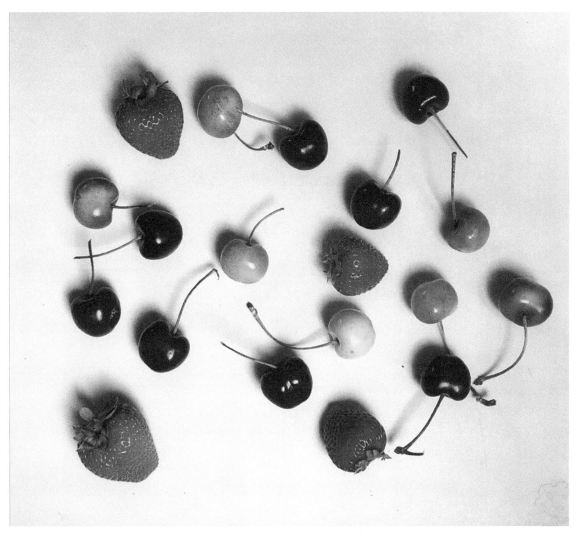

BIG IDEA 1

Building with and Talking about Shapes

If you are reading this introduction, I am going to assume that you love beautiful, creative mathematics and want to introduce this mathematics to your students. After many years of working in mathematics education, researching and teaching different groups of students, I have come to realize that there are two completely different versions of mathematics. One version is all that many people know; it is a subject of rules and procedures. But there is another mathematics—a subject of beauty and creativity, of openness and exploration—and when people encounter this mathematics, they are forever changed. That is because this mathematics illuminates our world; and when people learn it, they start to see the world in a new light. Paul Lockhart is a mathematician who has worked to communicate the real nature of mathematics to readers. He writes:

> What makes a mathematician is not technical skill or encyclopedic knowledge but insatiable curiosity and a desire for simple beauty. Just be yourself and go where you want to go. Instead of being tentative and fearing failure or confusion, try to embrace the awe and mystery of it all and joyfully make a mess. (Lockhart, 2012, p. 10)

I love this advice. Your students are young mathematicians, and you can introduce them to the mathematics that will capture their interest and curiosity if you share with them tasks that invite their thinking and their creativity. This big idea is the first opportunity for that lovely work. As a teacher myself, I am always excited to introduce my students to this version of mathematics.

This big idea is all about composing and decomposing shapes—or, to say it another way, putting shapes together and taking them apart. This is a mathematically worthwhile activity, as it helps students learn the qualities and characteristics of shapes. It also gives students opportunities to build and to move shapes around, helping them get to know their world and the role of mathematics within it.

In our Visualize activity, we give students the opportunity to study two-dimensional images of buildings and notice the shapes that make up the buildings. Students work to decompose the two-dimensional representations of three-dimensional buildings. They then get to design and build their own large city, paying attention to the shapes that make up the buildings. Students will be able to stroll through the streets of their city as if they were giants!

Our Play activity introduces a historical puzzle called the tangram. A tangram is made up of seven two-dimensional pieces that can be cut from a square. Students will be working out how they can fit their seven shapes into a frame, which will engage their minds in important mathematical thinking. Students will turn and flip their shapes as they try to solve the puzzles. We then ask students to make some of their own puzzles. Students get to create a design, name it, and the trace the outline of the shape for another student to try to build. Tangrams, comprising these seven simple pieces, have brought lots of enjoyment to people all over the world.

Our Investigate activity asks students to use tangram pieces to make different squares. We ask students to make as many squares as possible using their pieces, which will be an exciting challenge. Students have the opportunity to discuss shapes by name as they create squares of different sizes using the tangram pieces. When their collection of squares is complete, they talk about the different sizes of the shapes they have made.

Jo Boaler

Reference

Lockhart, P. (2012). *Measurement*. Cambridge, MA: Harvard University Press.

Build(ings)

Snapshot

Students explore how three-dimensional shapes can be composed into more complex figures by first constructing buildings out of recycled materials and then designing a community as a class.

Connection to CCSS
1.G.2, 1.G.1

Agenda

Activity	Time	Description/Prompt	Materials
Launch	10–15 min	Show students the Noticing Buildings sheet and ask what students think makes these buildings different or the same, then discuss what shapes are used to make the buildings. Support students in noticing different shapes. Connect these two-dimensional shapes to a simple building you have created with blocks, and discuss what shapes are used to make your building. Tell students about the activity and introduce them to the materials available for constructing buildings.	• Noticing Buildings sheet, to display • Simple building made from two blocks • Materials for building construction, to show
Explore	30–45 min	Partners make a plan for the type of building they want to create and the shapes they will need to construct it. They construct their building, making modifications to the materials as needed. Confer with students about their plans and the shapes they are using to compose their buildings, supporting the use of geometric language.	Materials for building construction, such as recycled boxes, masking tape, scissors, and markers, or wooden blocks

Activity	Time	Description/Prompt	Materials
Discuss	30–45 min	Do a gallery walk of the buildings and ask students to figure out what each building is and how they know. As a class, discuss how the buildings are similar and different, focusing on their components. Develop a class plan for how to assemble the buildings into a community, and then construct a village, town, or city, including streets. Discuss the components that appear frequently in buildings and those that do not, and develop ideas about the reasons.	• Students' buildings • Masking tape

To the Teacher

Building—composing complex figures out of simpler components—is critical for developing spatial reasoning and relationships, which in turn support all mathematical thinking. Composing and decomposing with shapes is strongly connected to composing and decomposing with numbers, and we want students to learn that mathematics involves seeing both the simpler components and the complex wholes they can create. What may look like play is far from trivial, and we encourage you to use this Visualize activity as a springboard to more opportunities to build in your classroom.

You can use a variety of materials for constructing buildings in this activity, including any wooden blocks you may already have in your classroom, but we find that using recycled materials is the most engaging and flexible. Ahead of this activity, invite students to collect and bring in boxes of different shapes and sizes to contribute to the project, including empty cereal, tissue, oatmeal, cracker, and shipping boxes. Students could also collect other forms, such as toilet paper or paper towel tubes. Students love working with recycled materials because they can create large buildings and, ultimately, a town they can walk through like giants. The recycled materials can also be altered by cutting doors, marking windows, and cutting rectangular solids into new shapes. Masking tape works well to attach the boxes, and students can tear the tape themselves.

To make this activity more locally relevant, consider expanding the launch by taking a photo of your own neighborhood or shopping district to show students in addition to the Noticing Buildings sheet we've provided. Students can analyze the shapes that compose local or familiar buildings, including their school. You may even want to take a walking tour around the block to notice and analyze the shapes used to compose buildings. Alternatively, you could conduct this walking tour in conjunction with the discussion of how to construct a town from buildings, noticing what kinds of building typically go together and how streets are used to make a community.

For the launch, you'll need to make a simple building out of two blocks, such as a triangular prism on top of a rectangular solid, that you can show students. This model will help students make connections between two-dimensional and three-dimensional space and notice how they are different.

Note that the discussion can include several different structures, including a gallery walk, whole-class discussion, small-group planning, and class construction of a town. Instead of your holding one long whole-class discussion, we encourage you to think of it in parts that include discussions of different sizes. If you include the construction of the town and possibly have students walk through it, this discussion portion could be an entire day unto itself.

Activity

Launch

Launch the activity by showing students the Noticing Buildings sheet on the document camera. Ask, What do you notice about these buildings? What makes them different or the same? Give students a chance to turn and talk to a partner about the buildings. Take some student ideas and allow students to make all different kinds of observations. Students may notice, for instance, that the buildings get used for different purposes, that they are different heights, or that some look familiar while others do not.

Then ask, What shapes make these buildings? Again, give students a chance to turn and talk to a partner. Invite students to come up and point to the shapes they see composing these buildings. Mark up the image to show the shapes students notice. Point out that different kinds of buildings use different kinds of shapes and that buildings can be made in many different shapes.

At this stage, students are likely to use two-dimensional language to describe what they see, such as *triangle* or *square*, and this makes sense with the image being shown. Tell students that while these images of buildings look flat, buildings in the world are made in three dimensions. Show students the building you built with blocks as an example of what it means to be three-dimensional. Point out that your building is solid, not flat, and so are buildings in the real world. Invite students to notice what shapes make your building, and provide language for these figures if students don't yet have words to describe them, such as *cube, rectangular solid,* or *triangular solid.* Tell students that they will be creating a building with a partner and that as a class you will be putting the buildings together to create a village, town, or city. Introduce students to the materials available for construction and the expectations for their use.

Explore

Provide partners with cardboard boxes or blocks to construct a building that can then be put together with others to compose a village, town, or city. If students are using recycled materials, they will also need access to scissors, masking tape, and markers. Partners discuss the following questions to create a plan:

- What do buildings look like?
- What kind of building might you build?
- What pieces will you need?
- How will you join the pieces together?

Encourage students to assemble and test different kinds of buildings, without attaching the pieces, before they settle on a plan for one building. As you circulate, conferring with students, ask questions to get them talking about the shapes they are using and the shapes' attributes or parts. Support students in using geometric language, such as *face* or *edge*, to describe what they are doing, joining, or planning.

Encourage students to add details to their buildings—for example, cutting doors, drawing windows, or labeling the buildings with signs—to make it clear what their building is.

Discuss

Hold a gallery walk of the buildings students have constructed. As students walk, ask them to think about the following questions:

- Can you figure out what each building is? What clues helped you?
- How are the buildings similar? How are they different?
- Which buildings would go near one another in a town or city? Why?

Gather as a class to discuss students' observations about the individual buildings and how they compare to one another. Ask, What buildings did you find most interesting or surprising? Why? As a class, discuss the question, How could we put our buildings together to make a village, town, or city? Develop a plan for assembling the buildings into a community. The class will need to consider where to put roads, what buildings go together and why, and how to make use of both sides of the street. You may want to allow students some small-group discussion time to develop ideas they can share and debate in the whole group.

Compose your city. You may want to mark off roads using masking tape. If you have built a large city with cardboard buildings, consider giving students a chance to walk through the city and make observations. Then compare buildings and their components by discussing the following questions:

- Which buildings have something in common? What is it?
- What differences do you see among the shapes we've made?
- What parts (3-D figures) were most commonly used? Why do you think that is?
- What parts (3-D figures) don't appear as parts of our buildings? Why do you think that is?

Look-Fors

- **How are students thinking about composing and decomposing with shapes to make buildings?** As you talk with students about their buildings, ask questions about the shapes they are using to compose larger buildings. What shapes might make up a building? How do different shapes create parts of a building? How do shapes work together to make a whole building? Buildings often have parts that we describe by their function rather than their shape, such as a lobby, apartments, or garage. Support students in connecting these parts of their buildings with the three-dimensional figures that they are using to represent them. If students are struggling to think about how to build a particular kind of building, it may be because they need a visual reference to decompose. You can support students in decomposing the building they hope to construct by showing them images of that type of building from the Noticing Buildings sheet, books, or the internet. When students can see the building they want to construct, ask, What shapes make this building? How could you make this building using the materials we have?

- **How are students grappling with different components and sizes when trying to make figures fit together?** If you are using recycled materials to construct buildings, you likely have a collection of boxes of various sizes that, unlike a set of wooden blocks, don't coordinate with one another. Making a tower, for instance, might require several boxes of the same size, but students may not have access to multiple identical shapes, those with the same width, or shapes of the same scale. As you circulate, pay attention to how students are reasoning about the ways that different solids fit together, or do not, and how students are attempting to solve problems of alignment or fit. If their boxes do not fit together in the ways they want, ask students, What could you do to get a better fit? Is there a way you could make this box smaller, or is there a different box you could choose? Encourage students to think about trimming or modifying boxes to get the lengths or angles they want. You may not have, for example, a triangular solid, which students may want for a roof. Ask, Is there a way to turn one of the boxes we have into the shape you want? Noticing that even the boxes themselves can be decomposed into other forms is a useful form of conceptual learning.

- **What language are students using to describe their buildings and their parts?** While students do not need the most formal geometric names for figures, having language for shapes makes it easier to discuss and compare them. Listen for the language students are already using, and support students in trying out new terms, particularly when they struggle to find a name for something they are trying to describe. Students will benefit from words to describe the forms they come in contact with most often—*prisms* or *solids*. With one of these core terms, students can then apply the two-dimensional names they have to describe the shape of the base of these objects to come up with a method for describing many of the solids they come across: rectangular solid or prism, triangular solid or prism, and so on. You may also have some additional shapes in your materials that have specialized names, such as cylinders or cubes. As you listen to students, ask them to describe the shapes they have used to construct their buildings and provide names for these parts when they use less formal language (such as *box*).

Reflect

What shapes are used most often to construct buildings? Why? What could it look like to construct a building that used unusual shapes?

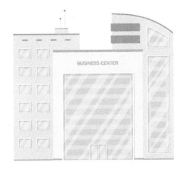

Solido:
Tridimencional
con volumen.

cubos
esferas
conos
prismas pirmide

Prisma:
es un solido que
tiene 2 buses
poralelas e
identicas unas rectans

Boaler, Je
Sources: 5-
84 all_is_

Tangram Puzzles

Snapshot

Students explore tangram puzzles as they develop ideas about composing and decomposing two-dimensional figures.

Connection to CCSS
1.G.2, 1.G.1

Agenda

Activity	Time	Description/Prompt	Materials
Launch	10 min	Show students the seven separate tangram pieces and ask students to make observations. Name each piece together. Tell students that this collection is called a tangram set, and provide a little history. Show students one of the tangram puzzles and explain that the goal is to use all seven pieces to compose the shape.	• Tangram set, to display • One of the tangram puzzles, to display
Play	25+ min	Partners choose a tangram puzzle to work on side by side, with each student having their own set of tangrams and copy of the puzzle sheet. Students explore how to compose the shape using the tangrams and outline the solutions they come up with.	• Tangram set, one per student • Make available: copies of all tangram puzzles, such that each student can complete multiple puzzles
Discuss	10–15 min	Discuss the strategies that students developed for solving the tangram puzzles, the clues in the figure outlines that helped them, and what made the puzzles challenging. Emphasize the language of composing and decomposing as you discuss students' thinking.	Tangram puzzles and tangram set, to display

Activity	Time	Description/Prompt	Materials
Extend	30+ min	Partnerships design their own tangram puzzle by creating a shape, outlining it, and giving it a title. Partnerships swap puzzles and try to solve. Discuss what made designing these puzzles challenging and how students had to think to compose a shape on their own.	• Tangram set, per student • Blank paper and marker • Optional: plastic sheet protectors and dry- or wet-erase markers

To the Teacher

In this Play activity, students explore ideas about composing and decomposing two-dimensional figures with tangrams, the classic dissection puzzle. Tangrams are considered a *dissection* because they are formed by dissecting, or decomposing a larger shape, in this case a square.

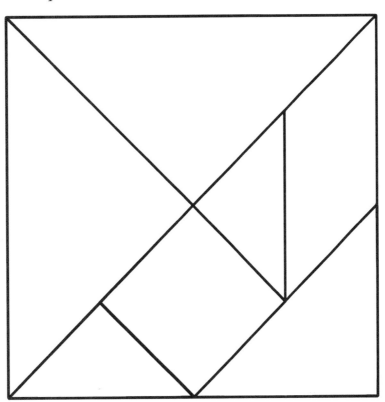

In this activity, we do not yet focus on the formation of the square—this will be at the heart of the Investigate activity. Rather, we familiarize students with tangrams through composing larger shapes in tangram puzzles. You can find literally thousands of these puzzles in books and online resources. We have provided six here, which are designed for students to overlay their pieces directly on the sheets. These puzzles fit tangrams that form a 4" square. We encourage you to explore other puzzles, particularly if you'd like to find those that connect with your students' interests or other class activities, such as read-alouds or science explorations. You can also use additional puzzles to turn this activity into an ongoing center or station.

If you decide to incorporate other puzzles, bear in mind that the presentation of the puzzle makes a difference in its complexity for students. Puzzles that are the same size as the pieces and that can be copied for children to overlay their pieces, as ours are, allow students to focus their cognitive effort on composing the figure. Cognitive effort is increased by removing either of these supports. For instance, if you present the puzzle on a document camera, students will need to work to transfer between the image and their work surface. Similarly, if students have a printed image but it is too small to overlay, they will need to transfer their thinking between the image and their work. We suggest waiting until students have lots of experience with these puzzle before introducing these challenges.

If you don't have access to tangrams sets, students can construct their own out of square pieces of paper. Provide students with 4" squares to match these puzzles, then lead them through the following directions to fold and cut pieces.

1. Start with a 4" × 4" square. Fold it in half along one diagonal. Cut along the diagonal making two large congruent triangles.

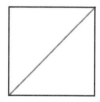

2. Cut one of the two large triangles in half.

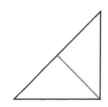

3. Take the other large triangle and fold the vertex of the right angle to the midpoint of the opposite side and cut along the line.

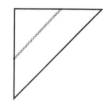

4. Put the two large and one medium triangles aside. Fold the remaining trapezoid in half and cut along the fold line.

5. Fold each trapezoid along the lines shown below and make a cut on the fold lines.

6. Now you have two large triangles, one medium triangle, two small triangles, one square and one parallelogram. The seven pieces make up the tangram set of shapes.

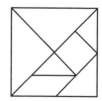

If you'd like to incorporate read-aloud into this lesson, there are several picture books that use tangrams. *Grandfather Tang's Story* by Ann Tompert is a classic, and you can use the tangram puzzles in the book as part of students' exploration.

Activity

Launch

Launch the activity by showing students the tangram pieces separately, rather than arranged in the square, on the document camera. Ask students, What do you notice about these shapes? Give students a chance to turn and talk about their observations. Discuss students' observations, descriptions, and names for these shapes. Be sure to come to agreement on a name for each shape.

Tell students that this is a famous collection of shapes called a *tangram set*, and tangrams are always composed of these seven shapes, sometimes called *tans*. Tangrams originated in China hundreds of years ago, though no one knows exactly when they were created or by whom. Tangrams are used for puzzles, which became very popular in Europe and America when they first arrived about 200 years ago. Show students one of the tangram puzzles they can work on today and tell them that in each puzzle, all seven pieces are used to make the shape, and the goal is to figure out how. Thousands of these puzzles have been created over hundreds of years, and today students will get to try out a few.

Play

Provide each student with a tangram set and access to tangram puzzles. Partners choose which puzzle to begin with, and each partner gets one copy of the puzzle. Partners explore the question, How can you use the set of tangram pieces to construct each shape? Students can work together on the same puzzle, but each must have their own space to explore, compose, and make adjustments. Encourage partners to sit side by side so that they can borrow each other's ideas and learn from each other's efforts. For each puzzle they solve, students outline their solutions on the puzzle sheet.

Discuss

Discuss the following questions as a class:

- What strategies did you use to solve these puzzles?
- What clues did you notice in the puzzles that helped you figure out where the pieces could go? (Show the puzzles that students want to discuss on the document camera, so that they can point out the features of the figure that supported them in finding a solution.)

- What made solving these challenging?
- What did you notice about the ways the pieces fit together (or didn't)? (Have the tangram pieces on the document camera so that students can show the ways they found the pieces fit together.)

As you discuss these puzzles, be sure to use the language of *composing* and *decomposing* shapes. Composing and decomposing, both with shapes and later with number, are big ideas that thread throughout first grade and beyond.

Extend

Invite students to create their own tangram puzzles. Partners design a figure, outline it on a blank piece of paper, and give it a title. Partnerships can swap their puzzles with those of other students and try to solve. If you'd like these puzzles to be used repeatedly, they can be placed in plastic sheet protectors so that the solution can be outlined with dry- or wet-erase markers and later erased. After students have designed their puzzles, discuss the following questions:

- What strategies did you use to design your puzzle?
- How did you have to think about the shapes to compose a larger figure?
- What made designing a figure challenging?

Look-Fors

- **Are students noticing shapes inside the tangram puzzles?** Some tangram puzzles provide clues about the ways pieces are used to construct them. Heads, feet, or other parts of the puzzles that stick out can point students toward particular pieces. By noticing these features, students can begin to decompose the whole into component parts, reducing the space that needs to be solved. If students are struggling to find an entry point for a puzzle, ask, Is there any part of this puzzle that gives you a clue about what piece might go there? Or, Can you see any shapes hidden inside this puzzle? Decomposing the tangram puzzles is a process, not a revelation, and needs to be tackled piece by piece.
- **Are students revising their solution ideas as they work?** As students work to solve these puzzles, it is useful if they recognize that each piece they put down may need to be moved. Some students may conceive of solving the puzzles as a series of choices (as in, "I'm going to put this piece here"), rather than as a process of testing and revising ideas toward a solution. Encourage students

to try out different pieces in different positions and to recognize that there may be multiple positions in which a particular piece *might* work. In this way, if students cannot find a solution with, for example, the largest triangles in a particular arrangement, then they would have something else to try next. Ask questions about possibility, such as, Where could this piece go? If this piece is here, where could these other pieces go? If this doesn't work, what could you change?

Reflect

What shapes in the tangram set fit well together? Show an example. Why do they fit together?

Reference

Tompert, A. (1997). *Grandfather Tang's story*. New York, NY: Dragonfly Books.

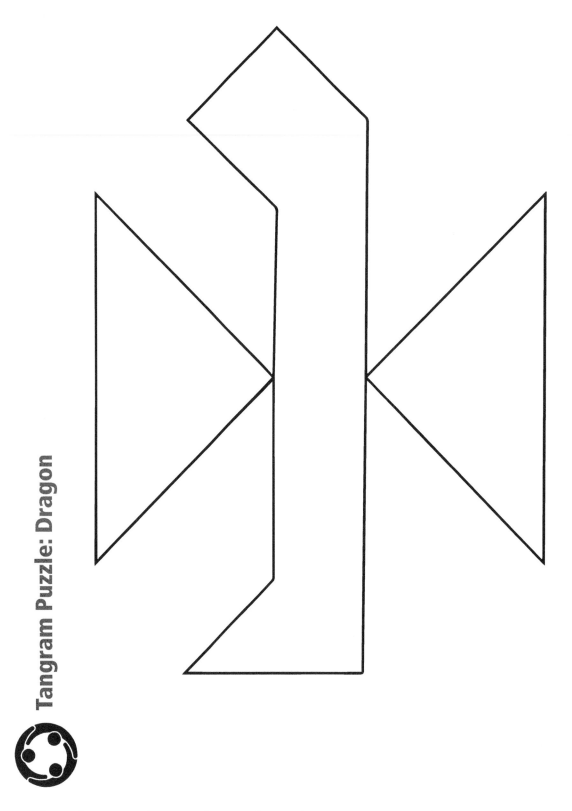

Tangram Puzzle: Dragon

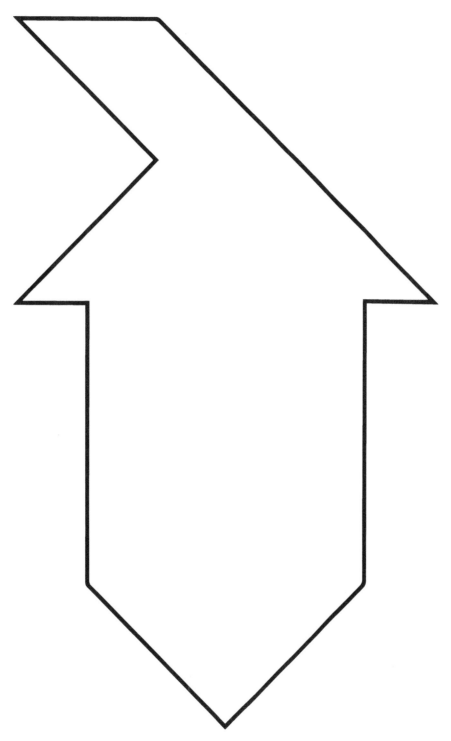

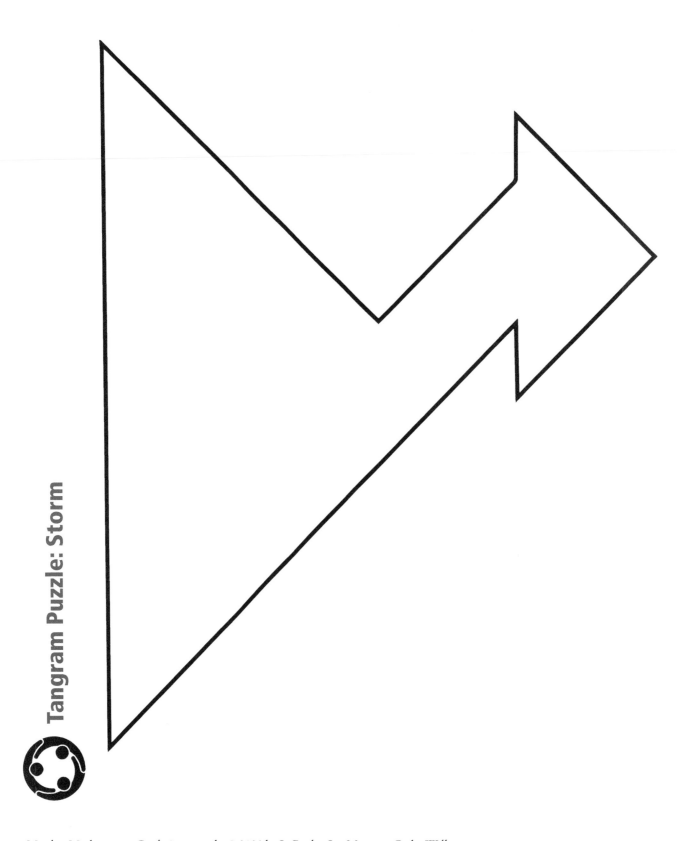

Tangram Puzzle: Flying Squirrel

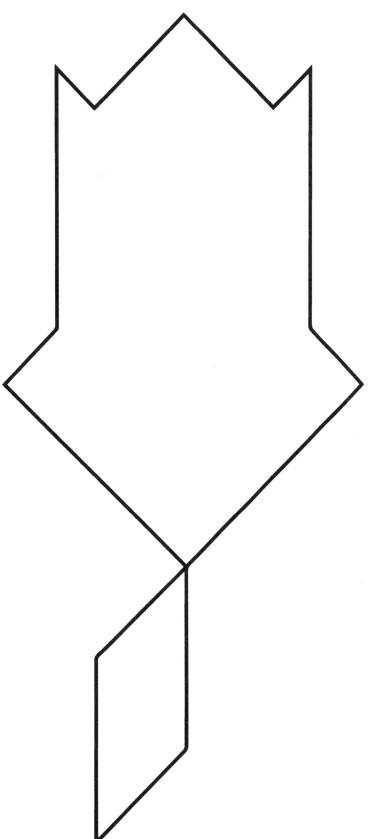

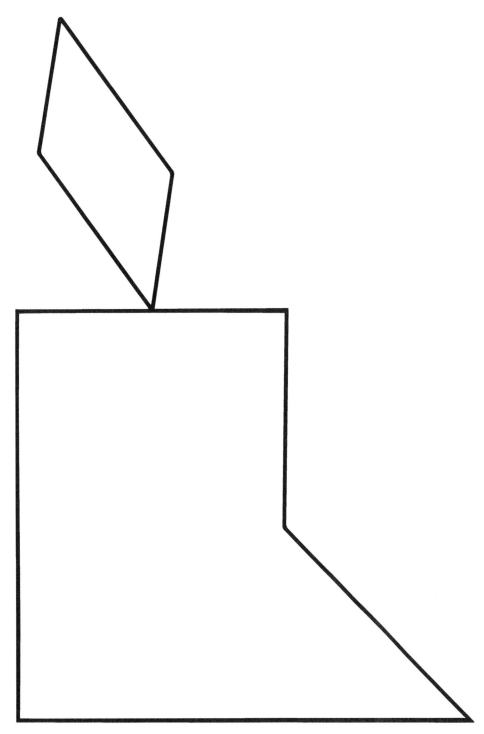

Tangram Squares

Snapshot

Students use tangrams to explore composing squares and investigate what makes two shapes or solutions the same or different.

Connection to CCSS
1.G.2, 1.G.1

Agenda

Activity	Time	Description/Prompt	Materials
Launch	10–15 min	Ask students what they know about squares, and after a turn and talk, record students' ideas on a chart. Support students in distinguishing between defining attributes and those that are not (such as color and size). Show students a set of separate tangram pieces and ask, What squares could you make with these pieces? Give students some time to think and visualize.	• Chart and markers • Tangram set, to display
Explore	20–30 min	Partners work together to generate as many squares using tangram pieces as they can, recording each on paper. Students investigate the question, What is the smallest square you can make? What is the largest square you can make? How do you know?	• Tangram set, per partnership • Tools for recording, such as half sheets of blank paper and rulers
Discuss	15 min	Invite students to share some of the squares they designed, then as a class figure out how to order the squares everyone made from smallest to largest in a display space. Looking at all the squares together, discuss whether there are different ways of constructing squares of the same size and whether the class has found all the possible squares.	• Display space and tools for posting students' squares

Activity	Time	Description/Prompt	Materials
Extend	30+ min	Provide partners with two sets of tangrams. Students investigate what squares they can make with 14 pieces and how this changes the size of the squares and number of solutions.	• Two sets of tangrams, per partnership • Tools for recording, such as half sheets of blank paper and rulers (consider using a different color of paper for the second set of squares)

To the Teacher

This activity, adapted from a task developed by NRich (nrich.maths.org), invites students to explore how to use the shapes in a set of tangrams to compose squares of different sizes. Key to this exploration is a shared understanding of what a square is. Students are likely to have informal knowledge of squares, which they encounter regularly and which are often named by adults, but they may not have ever articulated clear criteria for what makes a square. We encourage you to spend time in the launch discussing what features a square always has so that students can move forward with the investigation with a shared definition.

As students explore, the question of what constitutes "the same" is going to come up, and we see the discussion as a space for the class to work together to define what *same* means. For instance, are two squares that are the same size always the same, even if they have been constructed with different pieces? Are two squares that use the same pieces always the same, or does the orientation of those pieces matter? We think there is reason to include all versions of any formation of a square in the class's collection of squares. But students may compose a square out of the same pieces only rotated or flipped. Discuss whether two such squares are the same or different, and in doing so, you will be foreshadowing complex mathematical ideas about congruence and geometric transformations, regardless of the decision the class comes to.

The class needs to decide, Are any of these squares "the same"?

Activity

Launch

Launch the activity by asking, What do we know about squares? Give students a chance to turn and talk to a partner. Discuss what students know about squares, such as that they have four sides, they have square corners (right angles), and their sides are the same length. On a chart, record these ideas and draw a square to point out each feature. To help students articulate some of the features that matter, you might say, "I want to draw a square. How do I do that?" As you discuss squares, be sure to support students in distinguishing between attributes that define squares and attributes that do not (such as color or size). Students may also need to specifically discuss rectangles and note that a square is a kind of a rectangle that has sides that are all the same. Be sure that the class has a shared understanding of what makes a square.

Show students the set of tangrams as separate pieces on the document camera. Ask, What squares can you make with these pieces? How could you make a square? Allow students a minute to imagine how they might use the pieces to construct a square, but do not discuss solutions at this point.

Explore

Provide partners with a set of tangrams and space to record their squares. Half sheets of paper will enable students to record each solution separately, which will make ordering them by size easier later. Students explore the following questions:

- What squares can you make using any of the tangram shapes?
- How can you record the square you made?
- How many different squares can you make?

- What is the smallest square you can make? How do you know?
- What is the largest square you can make? How do you know?

For each square students make, they record both the outline and the shapes used to make it. Students may want access to rulers to support them in recording their thinking.

Discuss

Invite students to share some of the squares they made. For each one that students share on the document camera, you might ask, Did anyone else make this same square in the same way? Discuss the following questions:

- Which shapes were most useful in making squares? Why?
- Which ones were hardest to use to make a square? Why?

As a class, try to order by size the squares that students made. Ask, How can we put these in order from smallest to largest? Work together to create a display of the squares students made. Looking at the collection, discuss the following questions:

- Are there different ways of constructing squares of the same size? What makes two squares "different" squares?
- Do you think these are all the possible squares that could be made with these pieces? Why or why not?

Be sure to dig into discussions of what makes two squares the same or different, and into conversations about systematicity in looking for other squares. These get at larger mathematical ideas around structure—both looking for structure and using it—that will continue to come up throughout students' mathematical lives.

Extend

Extend the investigation by expanding the set of shapes that students can use to make squares. Provide each partnership with two sets of tangrams, a total of 14 pieces. Students explore the same questions:

- What squares can you make now?
- What squares can you make with two sets of tangrams that you could not make with one set?

- How many different squares can you make?
- What is the smallest square you can make? How do you know?
- What is the largest square you can make? How do you know?

Add the new squares to your display and discuss how having more pieces changes the number and possible sizes of squares that can be made. You may want to provide students with different colored paper or markers for this extension so that when you add the new squares to your display, you can distinguish between those that were made with one and two sets of tangrams.

Look-Fors

- **Do students have a clear working definition of a square?** Before sending students off to work on this investigation, it is critical that they have a clear and shared working definition of a square, developed during the launch. Students may use informal language to describe a square, such as simply providing examples in the classroom. Use these examples to press the class to develop criteria for what makes a square. You can revise the question to say, "I want to draw a square. How do I do it?" and use students' directions as places to press for reasoning. For instance, a student might say, "Draw a line," and you can ask, "How do I know when to stop?" which can ultimately get students to articulate that the sides need to be the same length. You might deliberately draw lines in ways you know will not create a square—for example, making an acute angle or sides of two different lengths—and when students protest, again press for why. Another frequent confusion is that squares are not rectangles. If you hear students voice this, gently reframe their language. For instance, if you drew two sides with different lengths, students may say, "You can't do that. That will make a rectangle," then you can say, "A square is a kind of rectangle. What makes it different from this one?"
- **Are students using equivalence to generate new solutions?** This task is a great place for students to think systematically about how to generate new solutions using equivalence. Several of the pieces in a tangram set can be swapped out for other pieces that cover an equivalent area, making a new solution to the task. For instance, students might make a square using the two largest triangles. If they notice that one of the triangles is equivalent to two medium triangles, they can exchange one large for two medium triangles and create a new solution.

Do students use equivalence to substitute the large
triangle for two medium triangles?

Students are unlikely to have language to describe the relationships they
are working with, but the underlying idea of equivalence is a critical one in
first grade, whether we are thinking of shapes or number. If you notice students generating solutions this way, ask them questions about their strategy
and invite them to share in the discussion. To encourage students to think
about equivalence, you can ask students questions such as, How could you
change the square you made to make a new square?

- **How are students conceiving of size when comparing squares?** We have
  challenged students to create the smallest and largest squares possible and,
  ultimately, to order these squares by size. Our intent is to think about a
  square's size by its side lengths, such that squares with longer sides are larger.
  However, some students may be conceiving of square size based on the number of pieces used to construct it. In this way, the two squares discussed in the
  previous question would be considered different sizes, because one is made
  with two pieces and the other is made with three. As you circulate, ask questions about which squares are bigger. Remind students that the question is
  what makes a square larger, and to focus attention on the outline of the figure
  rather than its components when considering size. During the discussion,
  press students to think about what makes a square bigger.

Reflect

What questions are you wondering now about tangrams? How could you explore
your questions?

Tens and Ones Are Useful Ways to Organize

The second big idea of this book focuses on place value, a key conceptual idea in mathematics. Researchers who study the learning of place value highlight the importance of students learning to group sets of objects into 10s and then treat the groups of 10 as units (Hiebert & Werne, 1992; Fuson, 1988; Steffe & Cobb, 1988). They also recommend that students learn to use different representations for groups of 10, such as physical materials and written symbols, highlighting different aspects of the grouping structure. This helps students build connections that lead to a more coherent understanding of place value.

In our activities, students are invited to make 10s out of classroom objects such as books, cups, and crayons; the spots on dice as they play with dice; and pencils in boxes. Across the three activities, students can see groups of 10 in contrasting forms and keep recordings of their groups, all centering on the idea of place value. This will help them build vital connections and see that the idea of groups of numbers is useful across different situations.

In the Visualize activity, students are asked to count collections consisting of 20–100 items. Students are not told to use a particular approach or method; instead we ask students to start their collections, and teachers to assist them as needed. The assistance focuses on grouping items by 2s, 5s, or 10s, which are benchmark sizes for collections, as they will help students learn the numbers 2 and 5 in our base 10 system. We recommend using collections of things that students will find in the classroom, such as books, snap cubes, and crayons. Students are offered tools for organizing, such as cups, bowls, and containers. The lesson concludes with a vital discussion about how students organized their collections and how the organization helped them. Students present their ideas, and others get to see the strategies they used.

In our Play activity, students roll 20 dice and organize them to see how many dots there are. This advances the learning from our Visualize activity, giving students the challenge of sorting groups of dots. Games played with dice are useful for students to learn counting and place value, but only some students will have had experience with dice games, so this lesson helps to equalize students' learning. Students will learn what 5 looks like on a dice, which will be a useful representation for them moving forward. This task encourages students to play and make sense of groupings as they play.

In our Investigate activity, students are given boxes of pencils and invited to see that 10 pencils are packed inside each box. Students are given different index cards showing a different total of pencils and are asked to work out how many boxes they need and whether there will be any pencils left over. The index cards have different two-digit numbers. Students make a class chart showing the number of boxes and leftover pencils for each of the index card numbers and then they look for patterns in the table. Students should start to see counting by 10s as counting a complete box of 10 pencils. The activity can be extended to three-digit numbers on the index cards. This activity gives students an opportunity to do something they may not have done before—move from groups of 10 to understand numbers, rather than move from numbers to groups of 10. Doing and undoing are important mathematical acts.

Jo Boaler

References

Fuson, K. C. (1988). *Children's counting and concepts of numbers*. New York, NY: Springer-Verlag.

Hiebert, J., & Wearne, D. (1992, March). Links between teaching and learning place value with understanding in first grade. *Journal for Research in Mathematics Education, 23*(2), 98–122.

Steffe, L. P., & Cobb, P. (1988). *Construction of arithmetical meanings and strategies*. New York, NY: Springer-Verlag.

Organizing and Counting a Collection
Snapshot

Students explore the connections between counting, organizing, grouping, and place value by working with collections of classroom objects.

Connection to CCSS
1.NBT.1, 1.NBT.2

Agenda

Activity	Time	Description/Prompt	Materials
Launch	10 min	Show students a collection of classroom objects and tell them you want to figure out how many there are. Ask, How could you organize the objects so that we can see how many there are? Discuss some ideas and tell students that their job today is to create ways to organize and count collections of objects.	Collection of classroom objects, such as snap cubes, coins, or books
Explore	25–30 min	Partners work with a collection of classroom objects to explore how to organize and count the items. Support students in developing ways to organize that make counting easier, such as grouping by 2s, 5s, or 10s.	• Collections of 20–100 or more classroom objects, such as math manipulatives, coins, books, or crayons, at least one collection per partnership • Tools for organizing objects, such as bowls, cups, or containers
Discuss	10–15 min	Discuss the ways that students organized their collections and how grouping made counting easier.	Optional: photographs of students' organized collections

Activity	Time	Description/Prompt	Materials
Extend	30+ min	Provide partners with collections and tools for organizing, and tell them that organizing by 10s and 1s can be the most useful way to count. Invite them to try this and explore how 10s and 1s help them count.	• Collections of 30–100 or more classroom objects, such as math manipulatives, coins, books, or crayons, at least one collection per partnership • Tools for organizing objects, such as bowls, cups, or containers

To the Teacher

This activity draws from the extensive research on early childhood mathematical thinking conducted by Megan Franke, Elham Kazemi, Deborah Stipek, and colleagues, which you can find in the wonderful book *Choral Counting and Counting Collections* by Franke, Kazemi, and Turrou (2018). Ideas about counting and the ways that our number system is organized require lots and lots of work to develop. Students need many extended opportunities across time to count aloud and count concrete objects to connect the spoken number word with the quantity being counted, and to understand that these quantities themselves can be organized in useful ways that help us understand how many there really are. We highly recommend the detailed activities and reasoning provided by Franke et al. and in a great unit of study in the Contexts for Learning Mathematics series, *Organizing and Collecting* (Liu, Dolk, & Fosnot, 2008). We see this Visualize activity as merely opening a door to a much longer exploration of counting.

In first grade, one central idea in counting is that you can organize objects into groups that make them easier to count. These can be pairs, groups of 5, or ultimately groups of 10. Students at this age often explore lots of ways to organize, including some that older students would not see as helpful, such as groups of three. Learning what ways to organize are most helpful is just part of a long process of trying different ways and asking, Is this helping me count? In the end, after lots of work counting, the goal is for students to select ways of organizing that support thinking about place value, particularly 10s and 1s.

A student counted 46 buttons by putting groups of 10 in each paper cupcake liner.

For this activity, you'll need collections of classroom objects of various sizes, from 20 to over 100, that students can hold, move, group, and count. Objects can be math manipulatives, such as snap cubes, tiles, pattern blocks, tangrams, or chips, or other objects you have in the classroom in quantity, such as bins of books, crayons, pencils, markers, dice, paper clips, or coins. You'll need more collections than you have partnerships, so that as students count, there is always another collection for them to try out when done. Students' stamina for counting varies, and you'll want to match students to a collection that stretches their counting a bit and for which grouping would genuinely help counting.

Along with doing this activity, students will need some experience with choral counting by 2s, 5s, and 10s. We encourage you to count the collections that are organized by 10s (and 2s or 5s) chorally together in the discussion, and to look for other opportunities to practice choral counting throughout the day.

Activity

Launch

Launch the activity by showing students a collection of classroom objects, such as snap cubes, coins, or books. Tell students that you want to know how many objects there are in this collection. Ask, How could you organize the objects so that we can see how many there are? Give students a chance to turn and talk to a partner about strategies for organizing. Take some ideas from students' conversations and have at least one student show some possible ways to organize.

Point out that there are lots of ways to organize, and some help us to count more easily than others. Tell students that their job today is to create ways of organizing the objects in their collection to make it easier for them to count and for others to see how many there are.

Explore

Provide partners with collections to count and tools for organizing, such as bowls, cups, or containers. Collections should contain more than 20 items and could have more than 100, depending on each partnership's stamina for counting accurately. Using their collections, partners explore the following questions:

- How many objects are in the collection?
- How can you organize the objects to see and count how many there are?
- How does your organization help you count?
- How can you record on paper how you organized and counted the collection?

Encourage partners to work with multiple collections as they develop systems for organizing and counting. As you circulate, ask questions about how students have decided how to organize, and how those systems are supporting their counting. Help students think about how they can record their work through drawings and labels to share with others.

Discuss

As a class, discuss the follow questions:

- How did you organize your objects for counting?
- How did you use tools (such as bowls, cups, or containers) to help you organize?

- Once you had organized your objects, how did you count?
- Which ways to organize helped you count? Why?
- How did you and your partner come to agreement about how many objects were in your collection?

Invite students to share the ways they organized, using the representations they created or the objects themselves. Another way to help students share their ways of organizing is to take pictures of students' collections to display. You could photograph the collections before and after organizing to discuss how this way of organizing helps everyone to better see how many there are. During the discussion, draw attention to ways that students grouped and how these groups connect to counting, such as counting by 2s, 5s, or 10s.

Extend

Once students have had lots of opportunities to count and organize, explicitly encourage them to organize using 10s and 1s by telling them that this can be the most useful way to organize. Provide students with collections and tools for organizing and invite them to try organizing the objects in groups of 10s and 1s. Partners explore, and later the class can discuss, the following questions:

- How does organizing in groups of 10 help you count?
- How did you know that you had 10 in each group?
- How did you count the leftover objects?

Look-Fors

- **Are students organizing collections to see how many?** Organizing can mean a wide range of things. For instance, there are ways of organizing that help us keep track of which items we've counted and which we have not, such as moving objects from one pile to another, and then there are ways of organizing that help us see and count how many, such as making groups. At the beginning of this work, you'll want to look for students who are focusing just on the first of these, keeping track of what has been counted, and encourage them to think of ways they could organize the collection to see how many there are. It is likely that some students will need to see others' ways of grouping objects in the discussion before they try this themselves. Developing systems takes time, which is why we recommend spending many days on this activity.

- **What size groups are students making?** Notice the ways that students are making groups. Are they grouping consistently, so that each cluster contains the same number of objects? Be sure that students notice that grouping only makes sense as a strategy if all the groups are the same size. Even when groups are the same size, you will likely see students choosing groups without thinking about how these groups support counting. For instance, you will see students make groups of three because they are easy to see, but then struggle to count by threes. Allow students to discover these challenges and then say, "I notice counting these groups is hard. What other groups could you make that would be easier to count?" Encourage students to revise and test their plan and then reflect on whether their new way was easier, and why.

- **Are students shifting from counting in groups to counting by ones?** Counting most collections means dealing with a few objects at the end that did not make a complete group. If counting a collection of 48 objects by 5s, students will get to 45 and then have three objects left over. There is concerted cognitive effort in switching from counting by 5s at one moment to counting by 1s starting at 45. Sit with students as they count, and pay attention to this transition. Do students continue to count by 5s pointing to individual objects instead of groups? Do they get stuck not knowing how to count the remaining objects? Do they simply need a few seconds to make the transition? You can support students by allowing them to think and reminding them where they were in their count so that they can pick up the thread counting by 1s. If students seem stuck, you can point to the groups and say something like, "These are 45. So, what is this one?" as you point to the next individual object.

- **Do any students group by 10s?** If you notice any students naturally grouping by 10s, be sure to ask them questions about why they chose this way to organize and then invite them to share this method in the discussion. Ask the class to reflect on this particular method. You might ask, What do you notice about this way of organizing and counting? Draw students' attention to how efficiently they were able to count. Encourage others to give this method a try on subsequent days of this work.

Reflect

How does grouping objects make it easier to count them? How do you decide what size your groups will be?

References

Franke, M., Kazemi, E., & Turrou, A. C. (2018). *Choral counting and counting collections.* Portsmouth, NH: Stenhouse.

Liu, N., Dolk, M., & Fosnot, C. T. (2008). *Organizing and collecting: Number sense.* Portsmouth, NH: Heinemann.

How Many Dots?

Snapshot

Students explore making groups to count by playing the dice game How Many Dots?

Connection to CCSS
1.NBT.2, 1.NBT.1, 1.NBT.3, 1.OA.5

Agenda

Activity	Time	Description/Prompt	Materials
Launch	5–10 min	Roll the dice on the document camera, then ask students, How many dots are there? How could we organize these to see how many? Invite students to share ideas for how to move and group these dice to count the dots.	At least 10 dice
Play	20+ min	Partners play the game How Many Dots? by rolling a large collection of dice, organizing them to count the dots, and recording how many dots they rolled. Students explore ways to group, organize, and count the dots, and see how large a collection of dots they can roll. Take photos of students' strategies, if possible.	• At least 20 dice, per partnership • Sticky notes, stack per partnership • Optional: device for taking photographs of student work
Discuss	10–15 min	Discuss the strategies that students developed for organizing and counting the dots on the dice. Draw attention to grouping, particularly by 10s. If you photographed students' methods, share these and discuss how they organized and how these systems help us count.	Optional: photos of students' organizational systems, to display

BIG IDEA 2: TENS AND ONES ARE USEFUL WAYS TO ORGANIZE

Activity	Time	Description/Prompt	Materials
Extend	15–30 min	Partners play a variation of How Many Dots? in which they roll the dice once and then develop multiple ways to organize the dice and count the dots. Partners draw and label each method to show how they counted the dots. Students explore which ways to organize and count made the most sense and were easiest for them to count.	• At least 20 dice, per partnership • Tools for recording thinking

To the Teacher

For this activity, you will need a lot of standard six-sided dice. We recommend at least 20 dice per partnership, because this large number presses students to form groups, particularly of 10, to deal with the inevitably large number of dots. If you don't have enough dice for the entire class to play at once, consider making this a small-group activity or station that students can rotate through.

The goal of this game is to encourage students to use the grouping strategies they developed in the Visualize activity to count the dots on the dice. In that activity, students may have formed groups of different sizes, such as 2s, 5s, or 10s, but dice can productively constrain these possibilities. With as many as six dots on a die face, using groups of 2 or 5 is limiting, and making groups of 10 makes more sense. Because the dots are pregrouped, students can use subitizing and their developing knowledge of how numbers can be joined to help them form groups. Through this activity, they may begin to recognize patterns for making 10, particularly that two 5s make 10 and that 4 and 6 make 10. However, if students still find that they need to count dots by 1s or 2s, or they form groups of different sizes, we encourage you to allow students to explore these possibilities.

This is a game that your class can return to repeatedly over time. The class can even track the highest total roll each week. Incorporating the idea of comparison between the numbers of dots in each roll introduces new ideas about how we know when one number is larger than another. Expect that these conversations will generate some debate, particularly about how to prove which number is larger. Students will likely use the counting sequence as evidence, and may also access other tools in your classroom environment, such as a number line or hundred chart, to support their thinking.

Students rolled 10 dice and organize them into groups of known sums.

Activity

Launch

Launch the activity by rolling a collection of at least 10 dice on a document camera. Ask, How many dots are there? How could we organize these to see how many? Give students a chance to turn and talk to a partner about strategies for organizing the dice. Invite students to come up and show some possible ways of organizing. Highlight ways of forming equal groups, particularly groups of 10, to help count the dots. Make connections between organizing these dots and the organizing and counting of collections that students did in the Visualize activity.

Play

Provide each partnership with at least 20 dice and sticky notes for recording their count. Partners play the game How Many Dots? multiple times.

Game Directions

- Roll all of your collection of dice.
- Organize the dice to help you count all the dots.
- Count the dots and come to agreement about how many there are.
- Record the number of dots on a sticky note.

As partners play, they explore the following questions:

- How can you organize these dice to help you count?
- How might groups of 10 help you count the dots?
- What is the largest number of dots you rolled? How do you know?

If possible, consider taking photographs of the ways that students organized their dice to show to the class during the discussion.

Discuss

As a class, discuss the following questions:

- How did you organize your dice to see how many dots?
- How did your strategies change as you played?
- What patterns did you notice that helped you organize?
- How did you count the dots without counting each dot individually? What helped you?

Draw attention to making equal groups, particularly groups of 10, and how these helped students count the dots without counting each dot.

If you've taken pictures of students' organizations, you can display these and ask the following questions:

- How did the students organize their dice?
- How many dots are there?
- How did the organization help us count?
- What other ways could the dice be organized?

If students made groups of 10, ask, What are the different ways you found that you could make 10?

Extend

Provide partners with at least 20 dice and tools for recording their thinking. In this variation of How Many Dots? partners roll the dice once and then develop multiple ways to organize and count the same set of dice. For each way students develop, partners draw how they organized the dice and counted the dots. Partners explore the following questions:

- If you make groups in different ways, do you still end up with the same number of dots? Why or why not?
- Which ways of making groups made the most sense to you? Why?
- Which ways of organizing made counting easiest? Why?

Look-Fors

- **Are students organizing dice into groups of the same number of dots?** Some students may be tempted to sort the dice by number, making a pile of ones, a pile of twos, and so on. This can be a useful first step to making groups of the same size, but if students stop at sorting without making groups, they may then resort to simply counting dots by 1s. Support students in thinking through how they could organize the dice so that they are easier to count. Make connections between the Visualize activity and what students are doing in this game. You might ask, When you were organizing and counting collections, what ways worked for you? You might also ask, How could you make groups of dots that were easier to count?

- **Are students making groups of 10?** Students may try making groups of a variety of numbers of dots, such as 5s, 10s, or even 12s. But groups of 10 are particularly powerful for counting in our base 10 number system. If you notice students making groups of 10, be sure to ask them questions about why they chose this way of grouping and how it helps them count. Invite students to share this strategy with the class. You may want to photograph how the students organized a set of dice into 10s and have the class count the dots chorally so that others can feel how easy this way of organizing makes counting. If you see students making other kinds of groups, ask them questions about how the groups work with the dice and how they count their groups. Students may have inventive ways of tackling both of these questions with groups of other sizes.

- **Are students noticing patterns for making 10?** This game foreshadows concepts to come, including joining numbers, making 10, and equivalence. It is not necessary that students see patterns for making 10 yet, but some may begin to notice that two 5s make 10 or that 6 and 4 make 10. They may even see that these are equivalent; by moving a dot from one die to another, a pair of 5s becomes a 6 and a 4. Some students may extend this thinking to more than two dice, perhaps discovering that three 3s and a 1 also make 10. There is no need to apply symbols to this work yet; rather, plant the seed for thinking about the different ways in which 10 can be represented. Highlight that these groups always work, that whenever you roll a 6 and 4 it always makes 10. Students may need to count the dots individually or on fingers to confirm this, but over time—both in this game and across the year—they will come to trust some of the ways that numbers fit together.

Reflect

What patterns did you use to help you to count the dots?

Boxes of Pencils

Snapshot

Students investigate how to pack pencils into boxes of 10 and loose ones, building connections between groups of 10s and 1s and our number system.

Connection to CCSS
1.NBT.2, 1.NBT.1, 1.NBT.3

Agenda

Activity	Time	Description/Prompt	Materials
Launch	10 min	Show students a box of pencils and tell them that pencils are often packed in groups of 10. Ask students, If we knew how many pencils we had, how could we figure out how many packs of 10 and how many loose pencils we had? Listen for student understanding of "packs of 10" and "loose ones," and address any confusion.	Box of 10 pencils, to display
Explore	25–30 min	Provide partners with an index card with a number of pencils written on it and snap cubes. Partners figure out how many packs of 10 they can make and how many loose ones will be left over, and then draw their thinking. Partners add their thinking to a class chart and try other numbers, as time permits.	• Snap cubes, for each partnership • Index cards with a different two-digit number of pencils on each • Class chart for recording number of pencils, number of packs of 10, and number of loose ones • Optional: sentence strips

Activity	Time	Description/Prompt	Materials
Discuss	15 min	Discuss students' strategies and observations from packing pencils into 10s and 1s. Look at the class chart of students' findings and discuss the patterns students notice. Use these patterns to predict how many boxes of 10 and loose ones you could make with a new number of pencils.	Class chart of number of pencils, number of packs of 10, and number of loose ones
Extend	30+ min	Repeat this activity, exploring how many packs of 10 can be made and how many loose pencils are left when you have a three-digit quantity of pencils. Explore boundaries that students often struggle with, including 100 and 110. Students add their solutions to the class chart and discuss how these number fit with or change the patterns students have noticed.	• Snap cubes, for each partnership • Index cards with a different three-digit number of pencils on each • Class chart for recording number of pencils, number of packs of 10, and number of loose ones

To the Teacher

In this activity, the goal is for students to connect groups of 10s and 1s with our number system, engaging in the mathematical practice of making sense of structure. Students explore different numbers of pencils to figure out how many packs of 10 they can make and how many loose pencils will be left over. We encourage you to use snap or unifix cubes to model pencils, with packs being formed by joining cubes into sticks of 10 and loose pencils being represented by loose cubes. It can be a leap for students to represent one object (pencils) with a different object (cubes), and some students may benefit from using actual pencils (or markers or crayons) instead, perhaps grouped in cups or bound by rubber bands.

You'll need to prepare a set of index cards with numbers of pencils recorded on each. These should represent a wide range two-digit numbers, such as 26, 46, 55, 60, 73, 81, and 93. You'll want numbers with different digits for 10s and 1s, some that are multiples of 10, some that have the same number of 10s but different 1s (such as 63 and 67), and some with the same number of 1s but different 10s (such as 18 and 38).

This variation creates lots of room for students to see patterns and notice structure. For the extension, you'll want to push these numbers into the low three digits, such as 107, 116, 129, 147, 200, or 212. Consider students' capacity for counting and the number of manipulatives you have for students to model the task.

Before you begin the activity, set up a three-column chart for collecting data during the exploration, with columns for number of pencils, number of boxes of 10, and number of loose pencils. Students' solutions in this table will be used in the discussion to look for patterns and make connections to our number system.

Activity

Launch

Launch the activity by showing students a box that hold pencils and telling them that when you buy pencils, they often come in boxes of 10. Ask students, If we know how many pencils we have, how can we figure out how many packs of 10 pencils we can make and how many loose ones would be left over? Give students a chance to turn and talk to a partner. As you listen in for students' strategies, pay attention to whether students understand the idea of a "pack of 10" and what "loose ones" might be. If students seem to not yet understand the difference between a pencil and a pack of 10 pencils, show students your own pack and the pencils it contains to clarify that it is one pack and also 10 pencils.

Explore

Provide partners with snap cubes and an index card with a number of pencils written on it. Partners explore the following questions:

- How many boxes can you make?
- How many loose pencils are left over?
- How can you check that you have the same number of pencils as the number on your card?

For each card that partners investigate, partners draw how they have packed their pencils. Then students record on a class chart the number of pencils, number of boxes of 10, and number of loose pencils. Alternatively, if you have access to sentence strips, you can make the chart out of these, so that you can reorder or group the rows for different comparisons during the discussion.

After students have recorded their thinking, provide students with a new card to try. Students can also make up their own number that they would like to explore.

Discuss

As a class, discuss students' strategies and observations through the following questions:

- How did you figure out how many boxes of 10 and how many loose pencils you had?
- Did you notice any patterns as you worked?

Display the chart of students' solutions. Invite the class to look at the data and discuss the following questions:

- What patterns do you see?
- What do these patterns mean?
- Will these patterns always be true? Why or why not?

In this portion of the discussion, be sure to make connections between 10s and 1s, the total number of pencils, and our number system. The goal is for students to begin to see that, for example, in the number 27, the digit 2 actually means 20, or two packs of 10, and the 7 represents seven individual objects. Students will likely benefit from the term *digit* to describe the idea that a number like 27 is formed with two digits, 2 and 7.

If you have used sentence strips, consider reordering or grouping some of the rows to compare quantities, asking, for example, What is the difference between two consecutive numbers (or between two numbers that are different by 10)?

Ask students to think about the patterns they have noticed, and ask them to make a prediction: If we had ___ (choose a number not on your chart) pencils, how many boxes of 10 and how many loose ones would you predict we could make? Why? If students are uncertain or not fully convinced, you might invite students to work with their partner to test or develop predictions and then return to the carpet to discuss.

Extend

Repeat this activity, extending the number of pencils to three-digit numbers (such as 107, 116, 129, 147, 200, or 212). Students will need access to a large number of snap cubes to model packing with numbers this large. If you do not have access to enough cubes, consider doing this activity as a station. Students often struggle to think about crossing boundaries, such as from 99 to 100 or 109 to 110. Use this opportunity to explore and allow students to struggle to make sense of those boundaries.

After students have figured out how many packs of 10 and loose pencils they have, invite them to add their solution to your chart. After collecting several solutions, discuss the patterns that students see in the chart. Encourage students to see numbers like 116 as 11 tens and 6 ones, which will later help them make the connection between tens and hundreds.

Look-Fors

- **Are students making boxes of 10?** In the Visualize activity, students were encouraged to organize objects in whatever way made sense to them, and may have made groups of 2, 5, 10, or something else. As you observe students using cubes to model this task, pay attention to the groups they make to ensure that they understand the constraints of this task: to make packs of 10. Similarly, students can miscount, even when intending to make packs of 10. Ask students how they know they have 10 in each pack and invite them to count them aloud for you to check. If students have made sticks of 10, they can alternatively check these by holding them side by side to compare length.

- **Are students thinking about both units: individual pencils and boxes of 10?** One of the conceptual moves in this activity is thinking about units in two ways at once. For instance, a pack of 10 pencils is simultaneously 10 pencils and one pack. Even our language when we discuss place value can sometimes be puzzling, such as when we refer to "a" 10 or ask "how many 10s?" Here "10" becomes a unit while also representing 10 individual objects. Expect that students might struggle with both this idea and the ways we describe it. Be sure to clarify language when students describe their work to ensure that they understand this distinction between 10s and 1s. For instance, if students point to a stick of 10 cubes and count "One," pause them to ask, "One what?" We want students to see that this is one *pack*, not one *pencil*, and, in fact, when they count the objects, you would want to hear them say "10" instead.

- **Do any students make predictions before they make boxes of 10?** Some students at this stage may enter each task with a fresh slate, forming packs of 10 and loose ones to solve the problem. Others, you may notice, may begin to make predictions about the number of 10s they will need, or modify their earlier work to create a new solution. These attempts are the beginning of attending to structure. If you see partners receive a new card and say immediately how many packs of 10 they will need to make, ask questions to probe their reasoning, such as, How do you know? Or, What did you notice in the number that made you think that? Similarly, you may see some students get a new card and then adjust the sticks of 10 they have already made, either putting some away or determining that they need to make more. Ask questions about their reasoning, such as, How are you going to use what you've already made to help you with this number? Why did you put some away? How did you know that you needed to make more sticks of 10?

Reflect

What do the digits in a number mean?

Representing and Modeling Joining and Separating Situations

The mathematics of the world, including the mathematics in the jobs that our students will eventually do, is filled with modeling. Some of the most important work in today's professions includes looking for relationships, patterns, and regularities. Traditionally, elementary schools across the world have focused their teaching on arithmetic—with a lot of emphasis on four operations—but the world needs people who are flexible, creative thinkers who can engage in powerful mathematical processes such as constructing, representing, and predicting, as well as organizing and analyzing data. Modeling is the form of mathematics that is rich in these imperative ways of working, and is great learning for first-grade students.

Conrad Wolfram is a mathematician and leader of Wolfram-Alpha, one of the most important mathematical companies in the world. (Wolfram technology powers Siri and Alexa.) Wolfram (2020) has communicated widely about the need to prepare students differently in schools. In particular, he points out that modeling is at the center of mathematical work and that it follows a four-step process, as shown in the figure.

| 1 | **DEFINE** QUESTIONS | 2 | **ABSTRACT** TO COMPUTABLE FORM | 3 | **COMPUTE** ANSWERS | 4 | **INTERPRET** RESULTS |

Source: Wolfram (2020, p. 54).

He also points out that too much time is taken in school training students to execute the third step, computation, when the most important learning is in steps 1, 2, and 4, particularly as there is extensive technology that will be used for step 3. This four-step process is what is known as modeling, and should be at the heart of students' mathematical work at all grade levels.

In our Visualize activity, we ask students to engage in this process as they take a story situation and define a mathematical question. They then turn it into an abstract form and calculate the answer, before interpreting the result. The two types of problems we bring into the Visualize activity are join—result unknown and join—change unknown. In both situations, it is as relevant for students to work out what is being asked (steps 1 and 2) as it is to calculate the answer (step 3). As students work to define a question, visuals will be really helpful, and the act of thinking visually will help stimulate important brain connections.

In our Play activity, students develop their modeling further with six different problem types. Teachers read a story to students and ask them, What is happening in this story? Students are invited to retell the story in their own words and make sense of the language of the problem.

Our goal is to help students focus on the words and work on deciphering and comprehending the meaning. Teachers can ask students, How could we model this story using objects or drawings? This is a time for exploring and modeling, and making sense of the situation, instead of just applying operations. Students will be given different story cards. The goal should be to celebrate the different ways of interpreting and modeling, and the challenges that students encounter along the way.

The Investigate activity is based around a story that does not have numbers; it is about the class going to the library. It is nice to think mathematically about situations that are not quantified. Students are asked to notice and wonder and then model what they are thinking. The class can work together to generate and choose a mathematical question that could be asked of the story, and to identify the missing information needed to answer that question. This is vital modeling work that will also help develop students' willingness to apply a mathematical lens to their lives.

Jo Boaler

Reference

Wolfram, C. (2020). *The math(s) fix: An education blueprint for the AI age*. Champaign, IL: Wolfram Media, Incorporated.

Showing the Story

Snapshot

Students develop multiple ways to model a joining situation you create that is relevant to your class.

> **Connection to CCSS**
> 1.OA.1, 1.OA.5, 1.OA.6

Agenda

Activity	Time	Description/Prompt	Materials
Launch	10 min	Show and read aloud the problem you have created. Ask students, How could we show with objects or pictures what is happening in this story? Invite students to share some ideas and discuss the possible tools they could use to model the story.	Problem you have written (see To the Teacher), to display
Explore	20+ min	Partners develop multiple ways to represent and model the story using materials and drawings.	• Problem you have written (see To the Teacher), to display and/or copied for partnerships • Make available: multiple math manipulatives, such as snap cubes, square tiles, and chips; the everyday object used in the problem; and colors
Discuss	15 min	Discuss the models that students created, how the models show the story, how they could be used to solve, and how different models relate. Introduce the language of *model* and *represent* to describe this kind of showing.	

Activity	Time	Description/Prompt	Materials
Extend	30+ min	Provide students with a different problem and ask them to model it using some of the same ways they used for the first task and some new ways they heard in the discussion. Discuss which model made the most sense for students and which they found to be the most useful.	• Problem you have written (see To the Teacher), to display and/or copied for partnerships • Make available: multiple math manipulatives, such as snap cubes, square tiles, and chips; the everyday object used in the problem; and colors

To the Teacher

For this activity, which focuses on encouraging students to develop ways to represent problem situations, we invite you to write your own problem to give to students so that it is locally relevant. Your students will be more engaged in the task if the context you select means something in your community, draws on the names of your students or people they know, and involves objects or situations they see regularly. Drawing on the research described in *Children's Mathematics: Cognitively Guided Instruction* (Carpenter, Fennema, Franke, Levi, & Empson, 2015), which we highly recommend and will dig into more in the Play activity, the problem you write should be one of two types:

- Join—result unknown. This is the typical structure for join problems that you have encountered, in which two quantities are offered and students are asked to join them to find the total.
  For example: I have 6 apples. You give me 8 apples. How many apples do I have now?
- Join—change unknown. In this structure, a starting quantity and the total are given, and students must figure out how much has been added to the initial value to get the total.
  For example: I have 6 apples. You give me some more. Now I have 14 apples. How many apples did you give me?

Notice that these two different problems can involve the same situation and the same numbers, but they demand different kinds of thinking. Indeed, the join—change unknown type is much more challenging for most students, and it can be modeled either as a join (How many more to get 14?) or as a separate situation (If I take away 6 from 14, how many do I have?). For your task, choose one of these two problem types and a context that is meaningful to your students. Be sure that the objects involved are things that students are familiar with and can easily visualize. For many students, it will be useful for these objects to be things you have in your class, such as pencils, crayons, science tools, or books. It helps if they can also be easily drawn; balls are easier for students to represent on paper than bears, for instance. When you choose numbers, the total should be no more than 20.

During the activity, the primary focus should not be on the answer but on the many different ways that students develop to model and represent the situation. Students may use a variety of objects to show the situation you create, and they may draw the situation to represent the action and mathematics embedded in it. While there will be only one answer, there are a never-ending number of ways to model the situation, and it is these that you want to encourage and discuss. Although we have described this as a single activity, we highly recommend that this kind of work be ongoing, with different problems offered over multiple days, so that students' strategies can be influenced by one another's and evolve over time.

Activity

Launch

Launch the activity by showing students the problem situation you've created, written on a chart or document camera. Read the story aloud, and then ask, How could we show with objects or pictures what is happening in this story? You may want to caution students that you're not focused on solving this problem yet, just on how to *show* it. Give students a chance to turn and talk to a partner, and then take some student ideas. If students name some manipulatives they could use, be sure to make these available during the exploration. You might also ask, Are there other tools we could use to show the story?

Tell students that they might find the answer to the question while they explore today, but their real goal is to find different ways to show what is happening in the story, using objects, drawings, or numbers.

Explore

Provide access to the text of the story so that students can refer back to check that their model matches what is happening. You can do this either by keeping the chart or document camera display posted or by creating and copying a sheet for each partnership. Provide access to multiple manipulatives, such as snap cubes, square tiles, chips, and the object involved in the story, if possible. Students may want to make any drawings concrete, accurately representing the details of the situation, so be sure to give access to colors.

Partners explore the question, What are the different ways you can show the story with objects or pictures? Students explore the same problem repeatedly, trying to create multiple ways to show the story with objects or pictures. The focus is not on solving—though students may—but rather on the models they make. As you circulate, ask questions about how students are representing the story, how the different parts of their models connect back to the story, and how they are using them to solve.

Discuss

Discuss the different models and how they are connected, using the following questions:

- How did you show the story?
- What tools did you use?
- How do the different parts of your model connect to the different parts of the story?
- How did (or could) you use your model to solve the problem?
- How are our models similar? How are they different?

As students share different ways of showing the story, be sure to put their ways side by side on a chart, carpet, or document camera and make connections between the models. Introduce the language *model* and *represent* during this discussion as ways we show the story to help us understand what is happening.

Extend

Provide students with another story and invite them to try to model it in some of the same ways they did with the previous story and some new ways that they heard from their peers in the discussion. Discuss the following questions:

- Which models make the most sense to you?
- Which do you find the most useful? Why?

If you have written a different problem type, discuss how the way the story was written may have changed how students modeled it.

Look-Fors

- **Are students focused on the answer or on the model?** Some students may be tempted to work quickly with the numbers, mentally or on their fingers, and arrive at an answer. It is critical to shift their attention away from the answer and toward representing the situation. If students do solve quickly, focus them on how they did so, by asking questions such as, What did you do with your fingers? Why? What did your fingers represent? What part of the story are you showing on this hand? What did you see in your mind as you were solving? How could you show that so that I can see it? When students can show or draw their ways of solving, then challenge them to develop new ways, emphasizing that creating different possibilities is the goal.

- **Are students able to connect their models back to the story, thinking about representation?** It is a big conceptual move to translate a story into objects, fingers, or drawings. But it is another big leap to connect those representations back to the story, to retain what each part represents and what it ultimately means for solving the problem. As you talk with students about their representations, focus attention on what the objects and the drawings they produce mean, and use the units involved in your story to label these aloud. For instance, if your story involves pencils in the bottom of a backpack and students have modeled this with square tiles, when you point to tiles call them "pencils" or ask, "Where are the pencils in the bottom of the backpack?" Connecting the story language to the representation is a key way to maintain the meaning of the model.

- **Do students' models represent the action or mathematics of the story?** Students may quickly represent the quantities in the story, and this is a solid first step in many problems. However, the mathematics of the problem is typically in the act of the story. For instance, a join problem, regardless of type, often involved verbs like *gave, put together, came,* or *found.* To solve the problem, students need to model the action in relation to the quantities, perhaps putting two groups together or taking them apart. As you look at students' representations, look for the action in your story and ask students to describe or show that action. You might say something like, "I see your nine pencils. Then what happens in the story? Can you show me?"

- **How do students transfer physical models to paper?** If students make physical models with manipulatives, eventually you will want to ask them to record these models on paper. We often say something like, "When we put these cubes back in the box, we'll lose all your thinking. Can you draw what you did on paper so that we can remember your thinking?" This move encourages students to extend and expand their models, essentially creating a written model of their physical model and developing connections between the story, objects, drawings, and numbers. Layering all of these representations together supports students in thinking flexibly about the mathematical concepts involved and about modeling as a practice. When you ask students to try transferring their manipulative work onto paper, notice what parts they represent and how. Students may need different tools for working with paper, where action might need to be represented with arrows or in multiple drawings. Invite these ideas into the discussion to support the class in thinking about how to model in different media.

Reflect

What kinds of models would you like to try tomorrow to help you see the story? Why?

Reference

Carpenter, T. P., Fennema, E., Franke, M. L., Levi, L., & Empson, S. B. (2015). *Children's mathematics: Cognitively guided instruction*. Portsmouth, NH: Heinemann.

Playing with Problem Types

Snapshot

Students develop their thinking about representing and solving joining and separating stories by playing with six different problem types.

Connection to CCSS
1.OA.1, 1.OA.6, 1.OA.5, 1.OA.4

Agenda

Activity	Time	Description/Prompt	Materials
Launch	10–15 min	Read aloud the library book story. Ask, What is happening in this story? Invite students to retell the story in their own words and make sense of the language of the problem. Ask, How could we model this story using objects or drawings? Students turn and talk and then share ideas. Share with students the Story Cards they can choose from today and challenge them to draw representations of the manipulatives they use to solve.	Library Books sheet, to display, or the task written in a chart
Play	25+ min	Partners choose a Story Card and tools for modeling the story. Partners work together to make sense of what is happening in the story, how to represent it, and how their representations help them solve the problem. Students move through problems at their own pace, recording their thinking as they go.	• At least four different Story Cards, multiple copies cut apart, for partners to choose from, or ones you have created yourself (see To the Teacher) • Make available: manipulatives for solving, such as snap cubes and square tiles, and colors

Activity	Time	Description/Prompt	Materials
Discuss	10–15 min	Discuss the strategies that students used to make sense of, represent, and solve the various problems they tackled. Invite students to make sense of one another's thinking as they share strategies and challenges.	
Extend	Ongoing	Return to this activity repeatedly over several weeks, adjusting and adding to the Story Cards made available. Continue to discuss innovative strategies and conceptual challenges as they arise.	• At least four different Story Cards, multiple copies cut apart, for partners to choose from, or ones you have created yourself (see To the Teacher) • Make available: snap cubes and colors

To the Teacher

This activity builds on the work students did in the Visualize activity to model and represent joining situations. Through this work and the discussion, students will have seen that there are many different ways to model mathematical stories. In this activity, students can use these models and develop new ones as they tackle problems of increased variety. The researchers who developed Cognitively Guided Instruction (Carpenter et al., 2015) identified 11 distinct problem types for joining and separating situations, each of which involves thinking differently about what is happening and how you might go about solving the task. We highly recommend a deep dive into their book to examine all of these problem types and the student thinking that emerges. In our activity, we focus on six of these problem types, including the two we used in the Visualize activity, to give students experience with both joining and separating and to ensure that students see and think through both joining and separating in different kinds of situations. Here we reproduce a portion of the table provided by Carpenter et al., with some examples of each problem type explored in this activity.

As you look at these different problems, you will likely notice that the types in the left-most column are the most prevalent in classrooms. They are also the most straightforward for students to solve because they can be modeled in the order the text is written. Students can read each sentence and perform its action with manipulatives, ending with an answer to the question. However, as we move to the right across the table, modeling becomes much more challenging because the unknown comes not at the end but in the middle or beginning. While all of these involve the same quantities and the same relationship ($7 + 4 = 11$), they are not the same for students. Students need opportunities to make sense of and grapple with all of these types of problems (as well as the other five types offered by Carpenter et al.).

Joining and Separating Problem Types

Join—Result Unknown	Join—Change Unknown	Join—Start Unknown
Sarai had 7 markers. She found 4 more in her backpack. How many markers does she have altogether?	Sarai had 7 markers. She found some more in her backpack. Now she has 11 markers. How many markers did she find in her backpack?	Sarai had some markers. She found 4 more in her backpack. Now she has 11 markers. How many markers did she have to start with?
Separate—Result Unknown	Separate—Change Unknown	Separate—Start Unknown
Sarai had 11 markers. She lost 4 markers. How many markers does she have now?	Sarai had 11 markers. Then she lost some markers. Now she has 7 markers. How many markers did Sarai lose?	Sarai had some markers. Then she lost 4 markers. Now she has 7 markers. How many markers did Sarai have to start with?

Source: Adapted from Carpenter et al. (2015).

In this activity, we have provided cards with problems that align with all six of these types for students to try out over time. However, just as with the Visualize activity, we encourage you to write your own problems too that draw on objects, people, and situations familiar to and engaging for your students. You can use these problems to make connections to other areas of inquiry in science or social studies, particularly if students are excited by a particular unit of study in these disciplines. While we have presented this as a single Play activity, we expect that students will need weeks to explore different problem types, develop their models, learn to represent different situations, and make sense of one another's thinking.

A caution and opportunity about language: It is unfortunately common for teachers to teach students to use key words in the text of a problem to determine the operation to use to solve it, such as that *altogether* means add or that *left* means subtract. There are at least two dangers to this approach. First, these words do not always mean to add or subtract, and students will later find the word *altogether* appearing frequently in multiplicative tasks, too, muddying the waters further. Second, not all problems are exclusively addition or subtraction. It is often entirely sensible to solve a separate problem by adding on, for instance. Segregating the tasks by operation prevents students from thinking creatively about their solution pathways and restricts the tools they can use. Far better is to focus on the work that you do as a teacher of reading and to support students in comprehending these tiny stories. Ask questions such as, What is happening in this story? What do you see in your mind? How can we act it out? These moves toward comprehension and visualization are the seeds of modeling and representation.

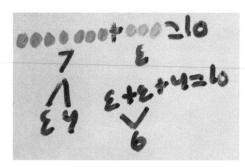
Student work for
"Some kids played in the park."

Student work for
"Paola had 11 library books."

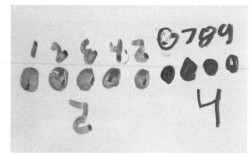
Student work for
"London has 5 crayons."

Sample student work for 3 of the activity cards

Activity

Launch

Launch the activity by showing students the Library Books sheet on the document camera, or the text of this task written on a chart. Read the task aloud, perhaps twice. Ask, What is happening in this story? Tell students that this is the first question they should always ask themselves when making sense of a math story. Give students a chance to turn and talk to a partner, and then invite students to retell the story in their own words. Be sure that multiple students have a chance to voice their way of thinking about the action of the story and that the class grapples with the idea that Paola returned *some* books, and we don't know how many. If students struggle to make sense of the story, invite some to act out what is happening using books from your own class. Point out that acting out a story is a way to make sense of it.

Remind students of the work they did in the Visualize activity to represent the story you wrote. Ask, How could we model this story using objects or drawings? Give students a chance to turn and talk to a partner. Then take some ideas about the tools they might use or how they might draw the action of the story. Tell students that today they will be able to choose the stories that they explore, represent, and solve. If your students are not yet independent readers, you'll want to read aloud the tasks available to help them choose. Tell students that if they use manipulatives, such as cubes, then after they solve, you'd like them to figure out how to record on paper what they did with the objects.

Play

Give students access to at least four different Story Cards from those provided or those you have created yourself, which represent a variety of problem types. Partners choose a problem to make sense of, model, and try to solve. Provide access to manipulatives, such as snap cubes and square tiles, and colors for representing thinking. Partners explore the following questions:

- What is happening in the story?
- How can you represent that story using objects or drawings?
- How can your model help you solve the problem?
- If you used objects, how can you represent on paper what you did?

Partners can solve multiple problem stories, moving at their own pace. Ask students questions about the tools they are choosing and how their models for one problem help them think about the next.

Discuss

As you circulate during the exploration, identify students to share interesting modeling strategies or challenges that would be worthy of sharing. This could include a new strategy, a way of drawing what students modeled with objects, number labels that students used, or a struggle to model that students encountered. Discuss these as a class using the following questions:

- What did the partnership do?
- How does their model match the story?
- How did you know what the action was in the story (joining or separating)?
- What questions do we have about what they did? Why does it make sense?
- How is it like anything you tried?

Be sure to point out where students are joining and separating, why, and that some problems can be solved in both ways. If students have encountered a challenge or an issue they couldn't resolve, bring that to the class and ask this question:

- How can we make sense of what is going on?

Extend

This play should be extended over weeks, and during that time, you can draw on new tasks from the Story Cards and write your own. We highly recommend that you read *Children's Mathematics: Cognitively Guided Instruction* (Carpenter et al., 2015) for additional guidance on how this work can unfold over time. During this extended effort, the most useful manipulatives for students to use to model these tasks are snap or unifix cubes. These can be built into sticks, joined, broken apart, and later made into sticks of 10 to reflect place value. Make these tools consistently available and place an emphasis on students sharing different ways they used the cubes to represent different kinds of situations and the actions that go with each.

Look-Fors

- **Are students interpreting the action of the story?** Accessing the mathematics in context starts with interpreting the story and being able to visualize the actors and objects involved and then what is happening to them. The action is often where a story becomes mathematized, moving from a fact (such as, "Paola had 11 library books") to a mathematical relationship unfolding in time. Ask students to retell the story to get a sense of how they understand the action, whether it is *loses, eats,* or *gives.* Then ask about how they have or could represent the situation and its action with objects or drawings: How could we show what is happening? Some students may find it more relatable to think of this as "acting out" the story with objects or with themselves as actors. All of these efforts support interpretation and representation.

- **Are students modeling problem types where the change or start is unknown?** When modeling situations in which the change or start is unknown, the story cannot be modeled linearly, as it is read. For instance, with the story, "Marcus has 12 rocks. His brother gives him 3 more. How many rocks does Marcus have now?" Students can represent Marcus's 12 rocks, then show someone giving him 3 more, and then use this model to count the answer to the question. All of this can be modeled sentence by sentence. However, with a task like the one in the launch, "Paola had 11 library books. She returned some books. Now she has 5 books. How many books did she return?" students may model Paola's 11 books and then not know what to do with the second sentence and have no idea how they have arrived at the third sentence. These are moments when discussing what is happening in the story can be useful, pointing out what we don't know, such as, "So Paola returned some books, but we don't know how many, do we?" You can ask questions such as, But what do we know? What are we trying to figure out? to help students locate the knowns and unknowns, which they might be able to model.

- **Can students connect the models they make back to the story?** When students create models, they are thinking abstractly about the story, transforming books and birds into cubes or circles. However, they can sometimes get stuck in this abstraction, forgetting that the cubes and circles represent real things that are interacting in the story in particular ways to produce some outcome. This can lead students to get stuck in the midst of problem solving, having represented some quantities but not yet knowing what to do with them. Or students might solve the problem and then arrive at an answer, such as 4, without a sense of 4 what or what that 4 tells us about the story. Ask questions along the way to support students in maintaining connections between their models and the story, such as, What does this represent? Why did you put these two groups together (or take them apart)? What happened in the story that told you to do that? What does your answer mean?

- **Can students narrate or explain the models they have created?** One of the mathematical practices you can be fostering in this work is explanation and justification. Students must actively learn to decompose a process into a sequence that they can share and to describe their reasoning for why they chose the particular method they shared. This is challenging, ongoing work for students in first grade and beyond. Give students opportunities to practice. You can ask questions to support their explanations, such as, Can you show me which blocks you're talking about? Then what did you do? Point out the ways they explain or justify that make their thinking clear and convincing, such as, "When you pointed to the blocks to show me Paola's 11 books, that really helped me follow your thinking." Sometimes you'll find that merely asking students to explain their thinking or models helps them identify and rectify errors independently. To grow students' capacity for engaging in this practice, be sure to constantly and consistently ask them to explain the meaning of the models they create to you and one another.

Reflect

What strategies help you make sense of a math story?

Reference

Carpenter, T. P., Fennema, E., Franke, M. L., Levi, L., & Empson, S. B. (2015). *Children's mathematics: Cognitively guided instruction*. Portsmouth, NH: Heinemann.

Paola had 11 library books. She returned some books. Now she has 5 books. How many books did she return?

 **Story Cards 1**

Marcus has 12 rocks. His brother gives him 3 more. How many rocks does Marcus have now?	Ash found 10 shells. He gave 6 away. How many shells does Ash have now?
London has 5 crayons. Kevin gives her some more. Now London has 9 crayons. How many crayons did Kevin give her?	Paola had 11 library books. She returned some books. Now she has 5 books. How many books did she return?
I have some books. I check out 6 more books from the library. Now I have 8 books. How many books did I start with?	Some kids played in the park. Then 3 kids went home. Now there are 7 kids in the park. How many kids were in the park to start with?

There are 4 pencils in the classroom cup.
Foa put 7 more pencils in the cup.
How many pencils are in the cup altogether?

Yusef had 13 crackers.
He ate 6 crackers.
How many crackers are left?

Kenya picked 3 flowers. How many more flowers does she need to pick to have 8 flowers?

There were 16 milks in the box.
Some kids drank milk.
Now there are 7 milks in the box.
How many milks did the kids drink?

Some birds were in a tree.
Then 9 more birds landed in the tree.
Now there are 13 birds in the tree.
How many birds were in the tree to start with?

Calvin had some dice.
He put 4 dice away.
Now he has 11 dice.
How many dice did Calvin have to start with?

Library Wonders

Snapshot

Using a notice and wonder routine, students explore a story about a visit to the library, generate questions, identify missing information, create models, and find solutions.

Connection to CCSS
1.OA.1, 1.OA.2, 1.OA.4, 1.OA.5, 1.OA.6

Agenda

Activity	Time	Description/Prompt	Materials
Launch	15 min	Share the Library Story and ask students, What do you notice? What do you wonder? Record students' observations and questions. As a class, generate and choose a mathematical question that could be asked of the story, and identify the missing information needed to answer that question.	Library Story, written on a large recording space
Explore	25–30 min	Partners choose what values they want to use for the missing information and create models to solve the problem. Students change these values, solving the problem repeatedly, each time creating a model and representing their thinking.	Snap cubes and recording tools, per partnership
Discuss	15 min	Discuss the decisions that partnerships made about the missing information, the models they created to solve, and the solutions they found. Create a class chart to record the information that students used and their solutions. Then, looking at the chart, invite students to notice and wonder about what it shows.	Chart and markers

Activity	Time	Description/Prompt	Materials
Extend	20–45+ min	Invite students to wonder mathematically about the library. Visit your school library or invite the librarian in to be interviewed. Generate mathematical questions, whether they can be answered or not. Consider using these to fuel new investigations.	Chart and markers

To the Teacher

This investigation draws on the I Notice, I Wonder routine, originally published by Max Ray (2013), who learned it from Annie Fetter, and which now appears on NCTM's website (https://www.nctm.org/Classroom-Resources/Problems-of-the-Week/I-Notice-I-Wonder/). In this structure, students are provided with a situation in which no question is posed and little information is provided. They work to make observations about the situation, pose questions, and then design a task the class can tackle in an open-ended way.

The story we've designed here centers on a visit to the library, because in many schools, students can leave the room to go to the library in small groups and sometimes all together, creating opportunities for decomposing the class into two smaller groups—those who go and those who stay. If there is a different space that your students visit flexibly during the day, you can adapt the story to better represent your context. We have deliberately made this story involve your actual class:

It's time to get new books. Some students from our class went to the library.

Students may ask questions such as: How many students went to the library? How many are left in the classroom? How many students are in the class altogether? You can use your own class to answer the final question, so that the total number matches your actual class, likely something between 18 and 30. Students can then decompose this total in many ways. For instance, if students decide that five students went to the library, then they can use your class size to determine how many are left. Conversely, they could decide that 12 students were left in the class, and then figure out how many went to the library.

This is merely one way this investigation could proceed. The students in your class may instead become much more intrigued by the number of books that were checked out, ignoring the number of students left in the class altogether. In this case, students might decide on how many students went to the library and how many books they each got, trying to find the total number of books checked out. This is one of the joys of this routine: students have agency to wonder about and solve the problems they design.

In the extension, we draw attention to the deeply mathematical nature of libraries. There is so much going on in these spaces, which we tend to think of as literary and information centers, that is inherently mathematical, and students have probably not thought about their library in a mathematical way. If you have a school librarian, inviting them into your classroom to share is a great way to encourage your students to think mathematically about a profession and wonder mathematically about their world.

Activity

Launch

Launch the activity by showing students the following story, recorded on a large surface, such as a whiteboard:

It's time to get new books. Some students from our class went to the library.

Ask, What do you notice? What do you wonder? Give students a chance to turn and talk. Then record their observations and questions, particularly those that are mathematical. If students have never used this routine, you may want to give them another chance to turn and talk to generate more observations and questions.

Then ask, What questions could we ask of this story? Give students a chance to turn and talk to a partner about the mathematical questions that they could pose of this story. Decide on a question to ask as a class, and record it where everyone can see it. For instance, one question that makes sense to explore is, How many students are left in the classroom? Student may come up with other questions that they find worth exploring. Once you have agreed on a question, identify the missing pieces of information, such as the number of kids out of the room, number still in the classroom, or the total number in our class.

Explore

Provide access to snap cubes for students to use to model the story, and tools for recording thinking. With a partner, students explore the following questions:

- What values (or numbers) could we use for the missing information?
- With this information, what is the answer to the class question?
- How could we model the situation?
- How could we record our thinking?
- What other values could we use? How do they change the solution?
- What other ways could we model the story?

Note that students should be able answer the question using not only multiple strategies but also different numbers.

Discuss

Invite students to share interesting ways of posing and solving the problem, using the following discussion questions:

- How did you begin to solve this problem? What decisions did you make?
- How did you model the story that you came up with?
- What models made the most sense to you? Why?
- Did anyone try anything new? How did it go?
- What different decisions did everyone make?

Make a chart to show the information from each partnership's decisions—for instance, with columns for number of students in our class, number of students still in the room, and number of students who went to the library.

After multiple partnerships have shared and you've added their solutions to the chart, invite students to look at the chart. Use the same questions that began this investigation, asking, What do you notice? What do you wonder? Discuss any patterns that students notice or new questions that the data inspires.

Extend

Invite your students to notice and wonder about the library, and use this as a place to invent more mathematical explorations. You might visit the library as a class to think mathematically, or invite your school librarian to come and be interviewed by the class. For instance, students may wonder about the following:

- How many books get checked out in a day or week?
- How many books get returned?
- How many students (or classes) come to the library each day?
- How often do some students (or classes) come?
- How many books does the library have?
- How many books does the library lose (or buy) each year?

Encourage students to wonder mathematically, whether or not they can answer their questions. Record students' questions on a chart. You may find some questions that the class or small groups can pursue in another exploration.

Look-Fors

- **Are students using a mathematical lens to think about the story?** During the launch, it is entirely appropriate that students contribute observations and questions that are both mathematical and not. Students may wonder what kinds of books the students will check out or who is going to the library. Honor these kinds of wonders alongside mathematical wonders too. However, if students are not posing mathematical observations or questions at all, you may want to invite them to do so by either explicitly asking for mathematical wonders or by modeling one of your own, such as, "I wonder, How many kids are going to the library?" This is likely all that students will need to bring their own mathematical lens to the story.

- **Are students choosing values for solving the problem that make sense?** Depending on the question that your class poses of the story and the missing information they will need to provide, some values will make sense, while others will not. For instance, if your class were to investigate how many students were left in the classroom, they would need to choose how many students went to the library. In such a situation, only values up to the size of your class make sense; there cannot be 64 or 112 students heading to the library from your room. Ask students questions about the missing information they are choosing and why those values could make sense. Be sure not to imply that the values are "right" in the sense that there is a single right answer; students can legitimately choose a range of values, any one of which could make sense. However, if you see students choosing values outside the sensible range, probe students with questions such as, Is this possible? Why or why not?

- **Are students connecting this work to previous work with modeling problems of different types?** In the previous activities in this big idea, students worked to develop ways to model joining and separating situations of different types. These models should be useful in this new context, and expect to see students drawing on the thinking they have been developing. For instance, if students have been using snap cubes to model how groups can become joined or separated, you can expect to see some students trying this for the library story. Students may also invent new ways of modeling and solving for this story, which should be welcomed as part of the long-term process of strategy development. However, if you see students becoming stuck, perhaps because the structure of this story is quite different and they need to provide some of the information themselves, ask questions that might help students connect this work to previous work. You might ask, How is this story like ones you have solved before? What models (or strategies or tools) helped you think about those stories? Would any of those help you here?

- **How are students organizing their thinking?** In this activity, students can solve the story repeatedly, choosing new values for the missing information each time. This means that students will likely need to record and organize their chosen information along with the solution it leads to and any work they have done. This is quite an organizational feat for first graders. As you circulate, encourage students to think about how they will record their work and how they can include the information they chose. For instance, students might take their decisions about how many students go to the library for granted and simply solve the problem with this information. You might point out that their work depended on that decision by saying something like, So this is your solution for when four kids go to the library, right? Can you write that part down so that we know how many kids were left when four kids went to the library? Students might add labels or write their decisions in a sentence. They could put each solution on a different paper or fold it to create boxes. Encourage them to think about this by asking, How will you keep each solution separate so that you won't get confused later?

Reflect

If you were to write your own math story, what would it be? What question would you ask? How would you solve it?

Reference

Ray, M. (2013). *Powerful problem solving: Activities for sense making with the mathematical practices*. Portsmouth, NH: Heinemann.

BIG IDEA 4

Using Data to Describe and Wonder about Our World

Recently I saw an interesting statistic: 90% of the world's data has been created in the last two years. This is worth some attention and thought, as it tells us that we are at a very crucial time in the world, one that should change the way we think about the mathematics we are teaching students. The world has suddenly and recently become filled with data—and our students need to learn to make sense of the data in their lives so that they can separate fact from fiction and become powerful users of data. This is not a change that affects just some of us; it affects us all. Each time we buy something online, travel through an airport, or visit the doctor, data is being used to inform decisions about our lives. Data literacy is an important goal for all of us, and data science is emerging as an vital new discipline. You may have had some experience of statistics in your education, but data science is a very different subject that starts with asking questions of data, making sense of data sets, and communicating findings through different visualizations. You can learn more about data science here: https://www.youcubed.org/data-science-lessons/. You can access a data science picture book that you may like to print for your students to color here: https://www.youcubed.org/resources/an-introduction-to-data-science-in-pictures/.

One of the exciting features of data science is its spirit of investigation—students learn how to ask questions of data, exploring situations that are meaningful to them. Students in first grade are not too young to begin their journey as data explorers.

One of the first and most significant ideas in the teaching of data literacy concerns the meaning of data. This is significant learning for students at any age,

107

and is often the first part of a college data science course. Our lessons start by asking students to make sense of categorical data, before they meet different data representations.

In our Visualize activity, students study collections of real-world objects to make sense of the world around them. Humans naturally seek to quantify information, and we use number and quantity to make sense of our world. Students are invited to make a collection of something from their world (e.g., leaves or flowers), learning that these collections of objects can be regarded as "data." Students classify their data and make a display. Students are intended to be learning to organize and describe their collections in a categorical way. This is a time for students to be creative in the ways they make sense of the attributes they notice and the ways they display the collection.

In the Play activity, students build from the previous lesson and focus on how they might display their findings. Students view an image of different types of seashells. They then look at a bar graph and a line plot that illustrate information about the shells. Students discuss and critique the examples, then move on to making similar displays for the collection they found and described during the Visualize activity. The goal is for students to be creative in how they display their data. The activity culminates in a gallery walk during which students look at the different interesting displays that students have produced. The reflection question in this lesson is important: "How did the data displays help you understand the collections?"

In our Investigate activity, students learn some ideas that are important to our environment. First, students look at an image of beach litter and discuss what they see in the image. Then students consider how trash can be sorted for recycling and study the beach image again. Ideally students will be given opportunities to sort trash from their school and make suggestions for ways to improve. This is the lovely opportunity that data science gives us—to investigate problems by using data and to suggest better ways to organize our world. Students collect, sort, study, and display what they collect. We hope that this set of lessons can culminate in an action that the class can take to improve their environment.

Jo Boaler

Organizing the Natural World
Snapshot

Students organize collections of real-world objects to explore connections between sorting and data displays.

Connection to CCSS
1.MD.4

Agenda

Activity	Time	Description/Prompt	Materials
Launch	10 min	Show students a collection of natural or real-world objects and ask, What makes the objects in this collection different from one another? Give students a chance to turn and talk and then collect some observations. Name these differences in color, shape, or other qualities as *attributes*. Tell students that they will be exploring and organizing a collection of objects using the objects' attributes.	Collection of natural or real-world objects, to display (see To the Teacher)
Explore	20+ min	Provide partners with a collection of objects and sticky notes for labeling groups. Partners explore the attributes of the objects and how they might group the objects to show what is in their collection. Students create a display of their data by organizing the objects so that others can see the groups and how many objects are in each.	• Collection of 10–20 natural or real-world objects, per partnership • Sticky notes, per partnership • Optional: device for taking digital photos of organized collections

Activity	Time	Description/Prompt	Materials
Discuss	10–15 min	Show some of the organized collections and invite the class to make sense of how they are organized. Students make observations and comparisons and pose questions of the data. Then discuss what ways of organizing helped the class understand the collections and make comparisons. Tell students that these objects are *data* and that by organizing, they made *data displays*.	Students' collections, to display
Extend	Ongoing	Turn this activity into a station. Place a collection of objects and sticky notes at the station and invite students to explore it and create a data display. If possible, photograph different ways of organizing the same collection to show students in a discussion. Ask, What do we learn about the collection from each way of sorting and displaying it?	• Collection of 10–20 natural or real-world objects • Sticky notes • Optional: device for taking digital photos of organized collections

To the Teacher

Data science begins with thinking about attributes and sorting the world into categories. In kindergarten, students gained experience sorting objects into categories, and in this activity, we connect the act of sorting to organizing a set of data. In order to represent and analyze data, we need to know what data is, how it can be organized, and what that organization can tell us. We begin with a set of objects as our data, making data something that is concrete, observable, and physically sortable.

For this activity, you will need to assemble collections of objects for each partnership to sort. These objects need to have multiple attributes and to be able to be sorted in multiple ways. We suggest that natural objects that are easy to collect in your area are well suited to this task; things like pebbles, leaves, or seashells all can be sorted in many different ways according to color, size, shape, structure, markings, and many other attributes. Alternatively, you can find ordinary classroom objects to sort, such as crayons, art supplies, or buttons. Avoid any objects, such as snap cubes, that have only one attribute that distinguishes one item from another, such as color. The collections you create for each partnership to organize should have 10–20 diverse items in each, though they should all be the same type of object. For instance, you might make a bowl of 15 pebbles, ensuring that the collection contains different colors, sizes, textures, and shapes.

The goal in this activity is for students to examine the objects and decide how to organize the collection into groups so that others can understand what each group has in common and how many objects are in the group. We have offered no mandates on the number of categories or what categories students develop, and we encourage you to leave this open ended so that students can explore what the various kinds of categories reveal about the collections. We are planting the seed for a very big idea: how you organize data determines what you can learn from it.

To launch the activity, you will need a collection of natural or classroom objects to share with students. The collection needs to be large enough that there is diversity in the pebbles, leaves, or seashells, but small enough that they can easily be displayed on your document camera. If the pieces in the collection overlap physically, the attributes of the objects will be hard for students to attend to. Alternatively, you can display the collection on the carpet if the objects themselves are large enough to be seen well by students.

Activity

Launch

Launch the activity by showing students, either on the document camera or the carpet, a collection of objects from the real world that you have assembled. Ask, What makes the objects in this collection different from one another? What are some of the differences you see? Give students a chance to turn and talk to a partner. Invite students to share some of what they notice, possibly about color, shape, size, decoration or markings, patterns, texture, or material. Point out that these features are called *attributes*, and attributes can help us compare and organize.

Tell students that they are going to be exploring a collection of materials like this to find a way to organize the objects into groups. The groups might have to do with the attributes they have already named or new attributes that they notice.

Explore

Provide partners with a collection of natural or real-world objects that could be sorted in a variety of ways, such as pebbles, seashells, leaves, or even a bucket of crayons or buttons or other art supply. Partners work to organize the "data" into groups so that others can see what makes the groups different and how many things are in each group. Provide partnerships with sticky notes for labeling their groups. Partners explore the following questions:

- What makes the objects in your collection different from one another?
- Which attributes do you think are important for showing differences in your collection?
- How can you show what is in your collection?
- What groups can you make? How would you describe each group?
- How can you make sure others can see how many are in each group?

Partners can try multiple ways of organizing the collection before settling on one way to group the objects. Partners then create a display of their collection, labeling each group they make and making sure others can see all the objects that belong in each group. If possible, take photos of these displays to show on the projector or smartboard during the discussion.

Discuss

Show different ways of organizing the collections, either by displaying photos of students' work or by moving some collections to the carpet for others to see. For each organized collection, ask the class to make sense of it using the following questions:

- How are these organized?
- What does this tell you about what was in this collection?
- What comparisons can we make between the groups?
- What questions does it make you wonder?

Be sure to examine collections that are organized in different ways, using different attributes, and that use different ways of arraying the items for counting. After you have looked at several ways to display a collection, ask, What ways of organizing made it easier to see the groups and how many there were in each group? Students may notice that labels, lining objects up, or placing groups side by side helps them understand the collection, count, and make comparisons. Tell students that one way of thinking about a collection is that it is a set of *data*, or information. These objects are information, and the way you organized the information into groups helps us learn about the objects; what students have made are *data displays*.

Extend

This activity can become a center where students explore different collections of objects and try to make displays in different ways. Place a collection of objects, similar to those used in the activity at the station, along with sticky notes. Students explore how to organize the objects into groups and create a data display. If possible, photograph these different ways of organizing the same collection over time. Then share multiple images of the same collection being sorted, organized, and labeled in different ways. As a class, discuss the question, What do we learn about the collection from each way of sorting and displaying it?

One collection of leaves sorted two ways. On the left *(clockwise from top left)*: round, miniscule, pointed, torn, and long. On the right *(clockwise from top left)*: short, light green, smooth, and bumpy.

Look-Fors

- **Are students attending to similarities and differences to make groups?** When looking at a collection of objects, students need to be thinking about both what makes the objects similar and what makes them different. On the one hand, if students become too focused on what makes them similar, they will end up with one large group they might call leaves, because they are, indeed, all leaves. In this case, you will need to ask questions to help students make distinctions, such as, What makes these leaves different from one another? Or, What are some ways that these leaves are different? On the other hand, if students become too focused on what makes the objects different, they might end up with 15 categories, each with one member, such as "brown, pointy, and large," "brown, smooth, and large," "pink speckled," and so on. In this case, you will need to ask questions that help students see similarities and create fewer, larger groups, such as, Which of these groups are similar? Or, How are these alike? What larger groups could we make so that we can see how the objects are similar?

- **Does every object belong to just one group?** Students are likely to make groups out of the objects in such a way that all objects belong to a group, but they may actually belong to more than one group. For instance, if students make groups for small leaves and pointed leaves, they may not consider that some leaves are pointed and small and simply place those leaves in the first group they form that applies. It is not necessary that students' groups be comprehensive or exclusive, but this is a worthy area for conversation. You might ask, How did you decide where to put this object? as you point to one that could go in more than one group. You might ask, Is this the only group it could have gone into? Just beginning to discuss this issue draws attention to the fact that objects each have multiple attributes and that each attribute could lead to a different way of categorizing it. If you notice any groups focusing on one attribute alone, so that the groups are truly exclusive, be sure to include this in the discussion.

- **How are students organizing the groups to see the data and make comparisons?** In addition to forming groups, students need to consider how to display each group so that others can see what the collection, and the groups, contain. Once students have groups, ask questions to help them develop a plan for displaying, such as, Is there an order these groups could go in? How can you organize the groups so that we can see how many are inside? How do you want to lay them out so that others can see what you found? How could you make it easy to count the objects? These aspects of data display are likely new for students, but they bear attention as we bridge student thinking from sorting to displaying data. Be sure to include a variety of displays in the discussion, to continue to support students' developing thinking.

Reflect

What ways of organizing a collection did you find most helpful? Why?

Displaying Data

Snapshot

Students play with representing their collections of natural or real-world objects using data displays, such as line plots or bar graphs.

Connection to CCSS
1.MD.4

Agenda

Activity	Time	Description/Prompt	Materials
Launch	10–15 min	Show students the From the Sea sheet and invite them to make observations about the collection. In turn, show students the Sea Shapes Line Plot and the Sea Shapes Bar Graph, and ask students how each shows the collection. Be sure students notice the features of each and make connections between the representations and the collection.	• From the Sea sheet, to display • Sea Shapes Line Plot, to display • Sea Shapes Bar Graph, to display • Optional: chart, tape, markers
Play	20–30 min	Partners revisit the collection of natural or real-world objects they explored in the Visualize activity to make groups and develop data displays of the information. Provide students with tools for drawing or assembling at least one data display. Using their display(s) and collection, partners create an information station that others can visit during the discussion.	• Collection of natural or real-world objects (see Visualize activity), per partnership • Make available: paper, sticky notes, 1″ grid paper (see appendix), rulers, and colors

Activity	Time	Description/Prompt	Materials
Discuss	15 min	Conduct a gallery walk and ask students to look for interesting ways to display data. As a class, discuss what students observed, including what surprised, interested, intrigued, and excited them about the displays they saw. Highlight clear and inventive data displays that students created.	Partners' information stations
Extend	20+ min	Provide students with books with various data displays and invite students to choose one and make sense of the data. Students explore what the display is about, the categories it shows, and what questions it raises. Students can then share these with the class or use them as inspiration for displaying data of their own.	Books with various data displays

To the Teacher

In this activity, we introduce a few ideas to bridge organizing collections into data displays. In the launch, we provide two different data displays—a line plot and a bar graph—showing the same collection, to give students a chance to make sense of these displays themselves. It may be helpful for students to have names for each type of data display to help them talk about their comparisons and the ways they are developing to display their own collections, but mastering these terms is not central to the lesson. What we do want students to understand is how their collections, and the ways they organize them, show *data*, or information. This data can be displayed in different ways to help others see what is in the collection and make comparisons.

We consider this a Play activity because we want students to play with and invent ways to display data about their collections. We suggest that you return to the same collections that students explored in the Visualize activity so that students are familiar with the attributes of these objects and can make decisions more readily about what categories they want to display. Be sure to encourage students to display the data creatively. They do not need to replicate the graphs shown in the launch or use their features in standard ways. The key is that students organize the data into some meaningful categories and can show how many objects belong to each category.

Whether or not you choose to try out the extension, consider making available books in your classroom that show different ways of displaying data. We love Steve Jenkins's (2016) book *Animals by the Numbers*, for instance. When data displays are used within text, they have a context that can intrigue students and support making sense of the data shown. Students may have seen these features in books before and not attended to them; point out how authors use data displays as another way of communicating ideas and information.

Activity

Launch

Launch the activity by showing students the From the Sea image and telling them that this is a collection of things that could be found at a beach. Give students a moment to look closely at what the image shows. You may want to tape this and the following sheets to a chart for easy comparison.

Then show the Sea Shapes Line Plot sheet, and tell students that this is another way to show the collection. Ask, What do you notice? How does this show the collection? Give students a chance to turn and talk to a partner. Then invite students to share what they notice about the line plot and how it is connected to the collection. Be sure students notice the labels that represent categories, the title, and how each square represents a shell or animal, and tell them that this representation is called a *line plot*. If you have taped this to your chart, you can label it Line Plot.

Show the Sea Shapes Bar Graph sheet and tell students that this is another way to show the collection. Ask, What do you notice? How does this show the collection? Give students a chance to turn and talk to a partner. Then invite students to share what they notice about the bar graph and how it is connected to the collection or the line plot. Again, be sure students notice the labels, the title, and how the grid is used to show how many are in each category. Tell students that this representation is called a *bar graph*, because of the bars used to show how many; if you have taped it to your chart, you can label it Bar Graph. Bar graphs and line plots are just two ways of representing how many things belong in different categories or information about a group, something we call *data*. You can give your chart a title such as Displaying Data and leave it up for reference during the activity.

Play

Provide partners with the same collections of natural or real-world objects that they explored and organized in the Visualize activity. Provide students access to paper, sticky notes, 1″ grid paper (see appendix), rulers, and colors. You could also offer die-cut shapes or stickers for students to use to construct data displays. Partners work together to create a way of displaying their data by exploring the following questions:

- What categories do you want to use to show the collection?
- How will you organize the data?
- How can you make a display of your data?
- What title makes sense for your display?
- What labels do you need?

Students can draw a display or construct one with sticky notes, die-cut paper, or stickers used to represent each object. Students may also choose to invent new ways of displaying the data. Be sure to keep this an open exploration of ways to organize information. Each partnership prepares at least one display and sets out their objects for others to see, creating an information station others can visit during the discussion.

Discuss

Conduct a gallery walk of the information stations each partnership has created, so that students can see the collection and the data displays side by side. As they walk around, ask students to be looking for interesting ways to organize or display the data.

Then, as a class, discuss the following questions:

- What did you learn from the data displays?
- What questions do you have about any of the graphs?
- What surprised you?
- What data displays did you enjoy looking at? Why? What did the authors of that display do?
- What did you see that you might like to try in displaying data in the future?
- How could you display your data differently? How might that help others see your data?

Tell students that there are many, many ways to display data, all of which have been invented by people who had data they wanted to share with others. Highlight any particularly clear or inventive ways your students have displayed their data.

Extend

Either as a whole-class or small-group activity, provide access to books that show different forms of data displays, such as those about animals, environments, communities, and people. Invite students to choose a data display from the text to make sense of. Students explore the following questions:

- What is this data display about? How do the title and labels help you?
- What kinds of categories does the display show? What makes these interesting?
- What do you notice about the display? What ideas does it show?
- What does the display make you wonder?

Students can then share this display with others to point out what makes it interesting, or use it to inspire displaying data of their own.

Look-Fors

- **Are students representing categories and counts?** Central to displaying categorical data like the kind that students are developing from their collections is communicating what categories they have used to organize the objects and how many objects fit into each category. It is not necessary that each object get counted exactly once. For instance, if students developed categories for "bumpy" and "spotted" and had a shell that was both, it is entirely appropriate for it to be represented twice, something that was not possible when students made physical groups in the Visualize activity. As you look at students' displays, look for these two features—categories and counts—and if you notice places where students have not indicated one or the other, ask questions to focus their attention. You might ask, What are each of these groups? How could you label them so that others will know? Or, How many rocks go in each of these categories? How could you show that?
- **Are students borrowing features from the data displays you shared?** The two data displays shared in the launch have a number of features that students might incorporate into their displays. Students might emulate the vertical and

horizontal structure used to organize information, use a grid, add number or word labels, or include a title. When you see these features, point them out and ask students where the idea came from and why they used it. The line plot and bar graph are intended as models, but it is critical that students use features not just to look like the model but because they help organize and communicate in particular ways. Focus conversation on what those features do to organize and communicate; students may not yet know, and it may be your role to point out how they helped you as a reader of the graph.

- **What new ways of displaying data are students creating that you want others to see?** While we have provided a pair of models of data displays, we hope that students come up with new ways to organize and represent data. They may try mash-ups of the two graphs, change superficial elements (for instance, the color of the bars or shapes in the line plot), or try something completely new. Pay attention for inventions, regardless of how successful they are, particularly where students can tell you what they were trying to do. Invite students to tell you all about their displays; point out inventive elements and ask students to describe what these features show. In the discussion, be sure to highlight features that help organize and communicate ideas, make comparisons or see the data, or remind you of other types of data displays not shared in this lesson. You may even want to find examples of other, related displays to share to show how others have used students' invented features.

Reflect

How did the data displays help you understand the collections?

Reference

Jenkins, S. (2016). *Animals by the numbers: A book of animal infographics*. New York, NY: Houghton Mifflin Harcourt.

Sea Shapes Line Plot

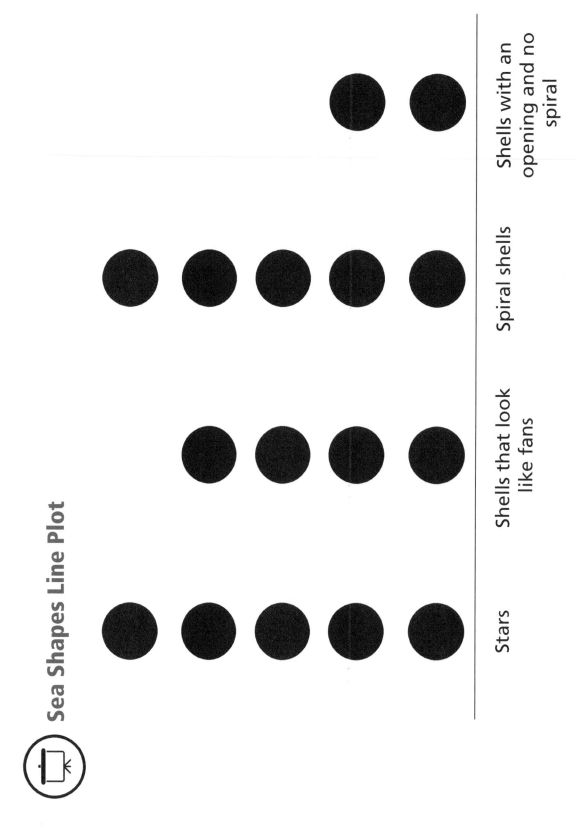

Stars

Shells that look like fans

Spiral shells

Shells with an opening and no spiral

Sea Shapes Bar Graph

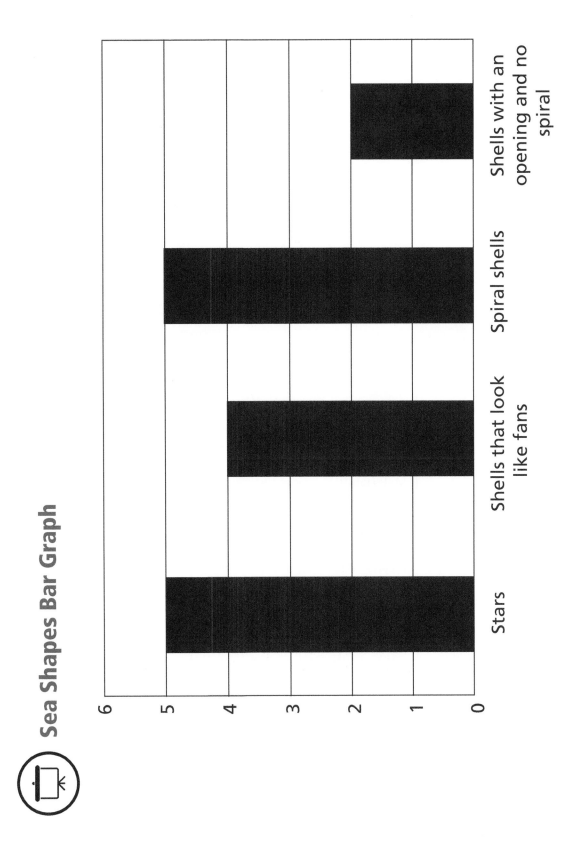

Our Trash, Ourselves

Snapshot

Students use a data lens to investigate what gets recycled in their classroom or school and to develop plans for taking action.

Connection to CCSS
1.MD.4, 1.NBT.1

Agenda

Activity	Time	Description/Prompt	Materials
Launch	15–20 min	Show students the Beach Litter image and tell them that trash damages our planet. Show students the Sorted Recycling image and tell them that all of these items can be recycled. Ask them to figure out how. Discuss materials that can be recycled. Return to the Beach Litter image to see whether students can find trash that could have been recycled. Ask, What do we recycle? As a class, develop a plan for investigating this question with data.	• Beach Litter image, to display • Sorted Recycling image, to display
Explore	Multiple days	Following the plan you developed in the launch, students work in groups or as a class to collect, sort, count, and record data about what gets recycled in your classroom or in the wider school. Students develop ways of displaying the data they collect.	• Gloves, per student • Recycling bin(s) and their contents, from your classroom and possibly others • Surfaces or containers for sanitary sorting • Make available: tools for creating data displays: 1″ grid paper (see appendix), paper, rulers, and colors • Optional: trash bin and contents

Activity	Time	Description/Prompt	Materials
Discuss	20+ min, across multiple days	At different points, discuss the ways that students are sorting, organizing, and counting their data. When students have data displays, share these, and you may also want to create class displays of data, combining data from across multiple groups. Discuss what they tell you about what gets recycled and what new questions the displays raise. Record these on a class chart. Generate ideas for action that the class could take in response to what the data shows.	• Chart paper and markers • Students' data displays
Extend	45+ min	Take action! Use the ideas students generated in the discussion to decide on a key action step. Provide students with materials needed and support them in implementing their plan.	Tools for taking action, such as construction paper and markers for making signs

To the Teacher

In this investigation, we position students to see data as a way of learning about themselves and their world, and to engage with data as something that can drive action. Trash and recycling are issues in every school, and ones that students have some control over, as they and their peers are often the ones creating and disposing of waste. Investigating what gets recycled, and possibly what gets thrown away, can be messy, both hygienically and mathematically. On the hygiene side, you'll need to think through what is practical and sanitary in your school context, whether it is sorting the recycling bins in your classroom and in others' classrooms, sorting trash, or even sorting the cafeteria trash. You'll need space to sort materials that you can adequately clean. This might mean covering tables with plastic or locating containers for sorting materials. Students will also need gloves before they touch waste. You may be able to get some from your school cafeteria or from science supplies.

This investigation will likely take several days; anticipate that students will need multiple rounds of data collection, organization, comparison, and display to make sense of what they are finding. In the launch, you will support the class in coming up with an initial plan, but as they work, they may naturally want to revise this plan to learn more. The plan they create in the launch does not need to be thorough, detailed, or final, but rather a starting point. This might include examining what gets recycled in your class over a series of days or comparing multiple classrooms or spaces. Students may start with your class or a neighboring room and then realize they need more data. Encourage this data-driven curiosity and provide time to explore students' questions whenever possible.

We have created space in the discussion and extension for students to develop plans for action based on their data. Depending on the data your class collects, what they find, and your resources, action could take a variety of forms. The most straightforward action might be to identify something that students themselves could be doing to support recycling in the school and design a poster campaign to promote this action, such as making sure recyclables end up in the recycling bins. Students could also design school announcements or promote action at a school assembly. Alternatively, students may want to advocate for changes that adults would implement, such as providing more recycling bins or getting recycling services at their school. They may want to encourage the school cafeteria to reduce waste by using fewer nonrecyclable materials. These action steps will involve preparing for conversation with decision-makers and using data to support their case. We encourage you to embrace as much as possible students' agency to effect change using data.

Activity

Launch

Launch the activity by showing students the Beach Litter image and telling them that Americans throw away more trash than any other people on earth. All of this trash damages our planet, and lots of it could have been recycled. If your students don't have much experience with recycling, we encourage you to spend some time discussing why we recycle and what it means to recycle. For instance, you can tell students that recycling helps us make use of materials again, so that they don't end up in a landfill.

Then show students the Sorted Recycling image. Tell students that all of these materials can be recycled. Ask, What types of materials do you see here? What categories have been created in this picture? Give students a chance to turn and talk to a partner about the materials and categories, then discuss students' ideas. Be sure students understand and notice materials such as glass, aluminum, paper, cardboard, and plastic. Tell students that all of these materials can be turned into something new if we recycle them.

Return to the Beach Litter image and ask, Do you see anything in this image that could have been recycled? Again, give students a chance to turn and talk to a partner, and then invite students to come up to point out items and name what material they are.

Pose the question, What do we recycle? If you decide to explore trash too, ask, What do we throw away? Are we throwing away anything that could have been recycled? Ask, How could we investigate these questions? As a class, develop a plan for how you might investigate what gets recycled (or thrown away) in your class or school. This plan should include where students will get their data, such as from your own recycling bin, those in other classrooms, or those in public spaces in the school. Students need to think about how they will sort, count, and record the data. You may decide to subdivide the tasks, with groups of students each investigating a different recycling bin, or you might decide to work collectively, sorting the contents of a bin as a class and then having each group count a category. There is no single way to tackle this investigation, and we encourage you to create time for students themselves to develop a plan.

Explore

As a class or in small groups, investigate your classroom (or school or lunchroom) recycling (and trash, if desired). Provide students with recycling materials as agreed on in the launch, surfaces (such as tables covered with plastic) or containers for sanitary sorting, gloves, and tools for constructing data displays, such as 1″ grid paper (see appendix), paper, rulers, and colors. Students explore the following questions:

- What do we recycle?
- What do we throw away?
- How can we organize the data to find out?
- How will we count the data?
- What gets thrown away that could have been recycled?
- What gets thrown away or recycled the most?
- How could we display our data?

Each group may collect its own data, depending on how you design the investigation as a class. If students have different data, you will also want the class to consider how to put their data together. For instance, if groups counted the recycling in different first-grade classrooms, they may want to add them together to find out what the first-grade collectively recycles.

Discuss

You may want to pull students together for several discussions along the way to support students in thinking about different parts of this investigation, such as sorting, counting, and displaying the data. This is also true if students continue the investigation across several days, collecting data in their classroom over a week, for instance. When you gather to discuss students' findings, use the following questions:

- What did we discover about what we recycle or throw away?
- How did you organize the data?
- How did you display the data?
- What ways of organizing and displaying helped us learn about ourselves or our school?
- What questions do our recycling and trash inspire?
- What action could we take based on the data?
- What additional data might we need?

On a Our Recycling Findings chart, record findings by listing the things that students have learned from their data. Create a Our Recycling Questions chart to track what you still do not know and what you might need in order to answer those questions. Depending on how your class decided to structure your investigation in the initial launch, students may decide that they need additional data to continue to explore these questions. This could include comparing data across classrooms, collecting data across the school, or focusing on one space, such as your classroom or the lunchroom, across multiple days or weeks.

As students gather data, their data displays may evolve. As a class, you may want to use the discussion to create displays together to communicate the data on chart paper and then consider what action your class could take, such as making signs to remind kids to put water bottles in the recycling or advocating for more recycling bins throughout the school. Record your options for action on a chart.

Extend

Take action! Using the data you collected, decide what action your class can take to promote recycling. For instance, if your class found that a lot of nonrecyclables end up in the recycling bin, then you may want to start a school-wide campaign to educate students on what can and cannot be recycled. Alternatively, if you find that students are not recycling something they could, such as packaging from the cafeteria, then students could design and post signs in the lunchroom to remind students to recycle these items at the end of lunch. Provide students with access to tools for taking the action they have chosen, such as construction paper and markers for making signs or an opportunity to talk with a school leader if they want to advocate for different policies or procedures.

Look-Fors

- **How are students counting the data?** One of the challenges of working with data like recyclables is that even counting itself can prove hard. For instance, students may want to count pieces of recycling, which makes lots of sense. However, if a piece of paper gets ripped in two, does the one piece of paper get counted as two pieces of recycling? What if the pieces are very tiny, as with wrapping that has been torn repeatedly? Is one piece of paper recycling the same as one piece of plastic? There are no correct answers to these counting questions, but interpreting the data will be easier if students make similar decisions about how to count. If you notice that different groups are counting differently or that questions bubble up about how to count, this would be a good place to pause the class for a brief discussion about what the questions are and how the class wants to resolve them as a group. If you develop any counting rules, you may want to record these on a chart for future reference.

- **How are students recording the data?** In the Play activity, students moved directly from sorting to displaying data. But in this investigation, students may want to collect counts before they create displays. This might involve recording data on paper, index cards, or sticky notes, or using new ways of writing down quantities, such as tally marks. Ask students, How are you keeping track of how many are in each of your categories? Or, How could you keep track? You could use this opportunity to hold another discussion to generate ideas for organizing count data, whether that is in a table or using tallies or snap cubes to represent each object. Although keeping track of data in a table is a conventional method, we encourage you to let students develop and explore different ways of organizing, so that they can find out the advantages and challenges of each.

- **What observations about the data are students making while sorting, counting, and recording?** As students sort and count their data, they will likely make casual observations to their partners or to you about what they see. Students may for instance say, "Wow! That's a lot of paper!" or "Only one pudding cup." These are important early observations about the data that could fuel additional questions or later interpretations. For instance, if students notice a lot of paper being recycled, you might wonder aloud, "I wonder why that is . . .," or if students notice the lone plastic pudding cup in the recycling bin, you might say, "Where are the rest, do you think?" or "Is that the only plastic anyone had at snack?" Students are unlikely to have answers for these questions, but the questions model being curious about the data all along the way and using that curiosity to fuel investigation. Students might then pose their own questions or decide to further explore the ones you posed. These observations might also be something you invite students to share in the discussion to find out whether other groups found something similar or whether the class can work together to figure out what the observation might mean.

- **What patterns in the data might point to action?** As you watch students sort and record their data, you might also have your own observations that could point to action. If you choose to take on the extension, you'll want to come up with some ideas about what the data is showing and what kinds of action that data might indicate to help you prepare for supporting student action. Taking action requires resources and planning, so your own observations of the data can help you generate ideas about what action is needed and what is possible in your school.

Reflect

What did data teach you about recycling in your school? How did data help you decide what action you could take?

Beach Litter

Sorted Recycling

Equal Means the Same

In this big idea, we help students understand an idea that they will use for many years in mathematics—equality and the equal sign. Different researchers have focused on the ways that students respond to the equal sign, to uncover why students have difficulties learning mathematics (Powell, 2012). They found something interesting. Many students have a fundamental misunderstanding about the equal sign (Sherman & Bisanz, 2009). The equal sign is what is known as a relational symbol; it indicates that the numbers on either side of the sign are in a relationship. The specific relationship they have is being equal. Wherever we see the sign, we expect to see equality on either side of the sign, and when students meet number sentences that have a missing number, such as $3 + \_\_ = 11$, their job is to find the number that makes both sides of the symbol equal. This may seem obvious to readers, but numerous studies have shown that students think the equal sign means "Do something" (Powell, 2012). They do not think it is relational; they think it is an operation, like adding or subtracting. This comes about when students are expected to solve problems with an equal sign without ever learning what it really means, and when they are given statements such as $3 + 2 = \_\_$. Without numbers on both sides of the expression, the sign does not convey a relationship but an operation.

When students go through years of mathematics thinking that the equal sign is an operation, they encounter problems in work with numbers, and later with algebra (Lindvall & Ibarra, 1980; McNeil & Alibali, 2005). Researchers have developed ways to teach the equal sign well, which involve both explaining to students the equality relationship (McNeil & Alibali, 2005; Powell & Fuchs, 2010; Rittle-Johnson & Alibali, 1999) and giving them many different ways of experiencing mathematical equality (Blanton & Kaput, 2005; Saenz-Ludlow & Walgamuth, 1998). The goal of this big idea is to give students the experiences that will enable them to develop a solid understanding of the relational nature of equality—through a theme of friendly monsters!

In our Visualize activity, students explore balance puzzles to help them build their understanding of the equal sign. The activity starts with a number talk in which students see a number on one side, with a few dots and a monster on the other. The question is, How many dots did the monster eat? Sometimes students will be asked to work out one number that satisfies the equality relationship. At other times they will find that more than one number can satisfy the relationship. The goal of the activity is less about finding the answer and more about encouraging students to justify their thinking and their strategies. Some students might count on, others might count back, and others might work with grouping. We encourage you to give students physical representations of the numbers, such as counters, bingo chips, and snap cubes. This will enable important brain connections to form, as students think about numbers symbolically and with visualization, touch, and movement. Students should be encouraged to construct a proof, an important mathematical act.

In our Play activity, students explore the meaning of equivalence by playing True or False? with equations. We again suggest a number talk format, in which students are given statements and asked whether they are true or false. This is an ideal time to reinforce the meaning of the equal sign and that we can only use it to represent equality. Give students time to think about each expression, poll the class to see who thinks it is true or false, and invite students to share their reasoning. Remember also to value incorrect thinking as much as correct thinking, because when students make mistakes, their brains are most active, and it is a positive moment for them.

Partners play True or False? by sorting a set of equation cards into the two categories. If students disagree or find an equation particularly challenging, have them bring it to the class to debate and come to consensus about. You may also want to share any good models or justifications that students created.

In our Investigate activity, students work out missing values by comparing numbers. This activity is not about finding an answer on the left and an answer on the right to match; it is about students explaining why their answer works, and being creative in their thinking, communicating, and justifying. These types of relational puzzles are important for students, and it is especially important that students are encouraged to consider the meaning of the statements. Too often students are given problems such as 100 – 98, and they set up an algorithm, as they are not thinking about what the statement is saying. Take students' answers and reasoning and highlight any nice examples of relational reasoning. If students do not share any, ask, Is there any way we could have figured out the missing number without finding the value of each side?

Jo Boaler

References

Blanton, M. L., & Kaput, J. J. (2005). Characterizing a classroom practice that promotes algebraic reasoning. *Journal for Research in Mathematics Education, 36,* 412–446.

Lindvall, C. M., & Ibarra C. G. (1980). Incorrect procedures used by primary grade pupils in solving open addition and subtraction sentences. *Journal for Research in Mathematics Education, 11,* 50–62.

McNeil, N. M., & Alibali, M. W. (2005). Why won't you change your mind? Knowledge of operational patterns hinders learning and performance on equations. *Child Development, 76,* 883–899.

Powell, S. R. (2012). Equations and the equal sign in elementary mathematics textbooks. *Elementary School Journal, 112,* 627–648. doi:10.1086/665009

Powell, S. R., & Fuchs, L. S. (2010). Contribution of equal-sign instruction beyond word-problem tutoring for third-grade students with mathematics difficulty. *Journal of Educational Psychology, 102,* 381–394.

Rittle-Johnson, B., & Alibali, M. W. (1999). Conceptual and procedural knowledge of mathematics: Does one lead to the other? *Journal of Educational Psychology, 91,* 175–189.

Saenz-Ludlow, A., & Walgamuth, C. (1998). Third graders' interpretations of equality and the equal symbol. *Educational Studies in Mathematics, 35,* 153–187.

Sherman, J., & Bisanz, J. (2009). Equivalence in symbolic and nonsymbolic contexts: Benefits of solving problems with manipulatives. *Journal of Educational Psychology, 101,* 88–100.

Hungry, Hungry Monsters!

Snapshot

Students explore balance puzzles to building understanding of the equal sign as showing two quantities with the same value.

> Connection to CCSS
> 1.OA.7, 1.OA.8, 1.OA.3, 1.OA.4, 1.OA.5

Agenda

Activity	Time	Description/Prompt	Materials
Launch	10+ min	Show students a Balance Number Talk and tell them that the equal sign means that the two sides have the same value. Ask, How many dots do you think are hidden? Give students think time, then a chance to turn and talk, before discussing students' reasoning. Ask students where the missing dots could be placed and why. Collect multiple possibilities. Draw and label each. Do a second number talk if there is time.	• One or more Balance Number Talk sheets, to display and record on • Marker
Explore	20–30 min	Partners work together to solve the balance puzzles, considering how they see the dots, the value of each side of the equation, how many dots are missing, and where to add them. Partners record their thinking by adding the missing dots to the puzzle.	• Balance Puzzle sheets • Colors and tinted bingo chips, per partnership • Make available: manipulatives for modeling, such as snap cubes or counters

Activity	Time	Description/Prompt	Materials
Discuss	10–15 min	Discuss the puzzle students solved, how they saw the dots, how they determined the value of each side of the equation and how many dots were missing, and how they decided where to place the dots.	• Balance Puzzle sheets, to display • Tinted bingo chips and markers
Extend	20–30 min	Partners design their own balance puzzles. Students consider what makes a puzzle interesting and how to design one that will balance when solved. After students have created their puzzles, partners can swap puzzles with another group to test out their designs.	• Monster Bank sheet, per partnership • Scissors, tape or glue stick, bingo chips, paper, and markers, per partnership • Optional: plastic sheet protectors and dry-erase markers

To the Teacher

In this activity, we encourage students to think in a different way about the equal sign than is commonly taught. Students often interpret this symbol as it is read aloud: "makes." But this leads students to think that *equal* means to execute an operation, like hitting the Enter key on an internet search. Instead, equal indicates not an action but a relationship. The two quantities on either side of the equal sign, regardless of their differences in appearance, have the same value. It is this relationship that we want students to understand deeply, so that they can think relationally about the quantities involved. To engage students in this kind of thinking, we have adapted Steve Wyborney's Splat dot talks (https://stevewyborney.com/2017/02/splat/) to involve balancing numbers, expressions, and dots across the equal sign.

In these puzzles, which we invite you to do as number talks in the launch and then have students try with a partner, the big idea is that the equal sign can be seen as a balance. Finding the value of one side tells you the value of the other side, which can then help you figure out what quantities are missing. We encourage you to ask students not just how many dots are missing and why, but where they could be placed. There is, of course, no one right answer to this question, but we want to encourage students to think visually about numbers, drawing on their knowledge of dice, the expression or quantity on the other side, rows and columns, and any other clues to see the full arrangement of dots in different ways. These arrangements support students in developing subitizing and seeing that there are numbers inside numbers, big ideas we began in kindergarten.

Activity

Launch

Launch the activity by showing the first number talk. Tell students that the equal sign in the center means that the value of the two sides is the same. Equal means the same, even when they do not look the same. The equal sign is like a balance, and the two sides need to balance. Tell students that some of the dots have been hidden (or even eaten) by the monster in the image. Ask, How many dots do you think are hidden (or eaten)? Give students a few seconds of think time, then ask them to turn and talk to a partner. Invite students to share their ideas. Be sure to ask, How many dots do you think are hidden (or eaten)? How do you know? Draw attention to thinking about the equal sign and how many need to be on each side to make them have the same value.

Ask, Where would you add the dots? Note that there is no one correct answer, but that students may see the number differently and imagine the dots in specific places. Encourage multiple ways of seeing the existing dots and the new dots. Record student thinking both visually and symbolically. For instance, after you have drawn the new dots where students wanted to place them, you might label that group as "6" or choose to label it "3 + 3" because there were three dots and students added three more. However you choose to label the dots symbolically, describe for students why your label makes sense. You can do a second number talk if you want students to have more experience, particularly with images that have an expression on one side of the equal sign.

Explore

Provide students with access to the Balance Puzzle sheets, colors, tinted bingo chips, and any other manipulatives they might want for modeling, such as snap cubes or counters. For each puzzle, partners work together to answer the following questions:

- How do you see the dots in the puzzle?
- What is the value of each side of the equation?
- How many dots are hidden by the monsters? How do you know?
- How can you balance the two sides so that they are equal?

- Where do you think the dots could be added?
- How can you record your thinking so that others can see what you see?

Partners can work on multiple puzzles, recording their thinking as they go. As you circulate, focus conversation on equality and the value of each side of the equation.

Discuss

Invite students to share how they thought about the puzzles they solved by discussing the following questions:

- How did you know how many dots were hidden (or eaten)? How did you see it?
- How did the equal sign help you?
- Where did you add the dots? Why?
- Did you and your partner think differently about any of the puzzles? How were you each thinking?
- Were any of the puzzles challenging? What questions do you have, or how did you work through the challenge?

As you discuss the different puzzles that partners solved, display each on the document camera and provide bingo chips for students to show how they placed the dots. Throughout the discussion, return to the big idea that the two sides of the equation must have the same value.

Extend

Provide students with the Monster Bank sheet, scissors, tape or glue stick, bingo chips, paper, and markers, and ask partners to design their own balance puzzles. Partners can cut out the monsters from the Monster Bank to make part of their puzzles. Partners use the following questions to support their thinking:

- How will you make sure the puzzle can be solved?
- How can you design an interesting puzzle?
- How will you make sure it can balance?

Partners can swap with others to solve new puzzles. If you want students to be able to solve them repeatedly without marking on the original puzzle, you can either ask students just to use bingo chips without marking the paper or place each puzzle in a plastic sheet protector and provide dry-erase markers.

Look-Fors

- **Are students thinking about the value of each side of the equation?** The big idea is that equivalence means that two things have the same value. For instance, 5 + 3 has the same value as the number 8 and as a collection of 8 dots, or, for that matter, as 10 − 2. Each of these has a value of 8, and the equal sign indicates this relationship. When you observe students in the number talk and during the exploration, look for indications that students are thinking about value and equivalence. Looking at 5 + 3, students may say things like, "That's 8," or "That means we need eight dots over here." Use these as opportunities to ask questions and revoice to press this idea that 5 + 3 has the same value as 8, that these are equal. Alternatively, you might see some students thinking of the equation as a sentence that is read in order, which can lead to interpretations that do not use equivalence. For instance, in the number talk that involves 5 + 3, students might read the dots on the left first as "6 equals 5" and then become stuck, because 6 does not equal 5, and no matter how many dots are added, they cannot make this equation true. Point out that the entire side of the equation matters and ask what the value of the side is, helping students think about each side in turn rather than as a series of steps read left to right.

- **How are students thinking about the number of missing dots?** Figuring out the number of missing dots is at its heart a "how many more" problem, as in "I have six dots. How many more do I need to equal eight?" Students might think about solving this problem as a missing-addend problem, as in 6 + __ = 8. Alternatively, students might solve this by counting or counting on. For instance, students might count all the dots to 6 and then add bingo chips for 7 and 8, or simply start with 6 and count on "7, 8" with fingers to see that they need two more dots. Students might also conceive of this is a separate situation, as in 8 − 6 = __. Each of these different methods makes different connections between joining, separating, measurement, and counting, and you'll want to highlight that these different ways of thinking about the missing dots are connected and are themselves equivalent.

- **How are students modeling the parts of the puzzle?** Because each puzzle includes both images and symbols, students need to think through how they want to model the different parts to think about equivalence. Some students might find that using bingo chips for both sides makes sense because that enables them to see more easily when the two sides are the same. However, students may have more experience modeling with snap cubes at this point, and if they choose these, they will need to think through how to relate the cubes on one side to the dots on the other. Students may also choose to transform the dots into cubes to address this conversion issue, but they will then need to translate cubes back into dots when they add on the missing dots. Note how complex this work is, moving between different forms and representations while grappling with equivalence. As you circulate talking with students, continue to emphasize what parts of the puzzle are equivalent. For instance, modeling 5 + 3 with five red cubes joined to three blue cubes, you can point out that the red cubes are equivalent to the 5, the blue cubes are equivalent to the 3, and the joined stick is equivalent to both 5 + 3 and the value on the other side of the equation.

Reflect

What does the equal sign mean? Give an example to help you explain.

$$6 =$$

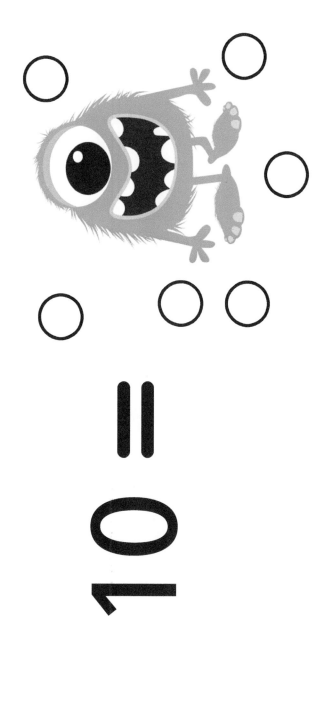

$10 =$

$$= 5 + 3$$

$$8 - 1 =$$

$$= 15$$

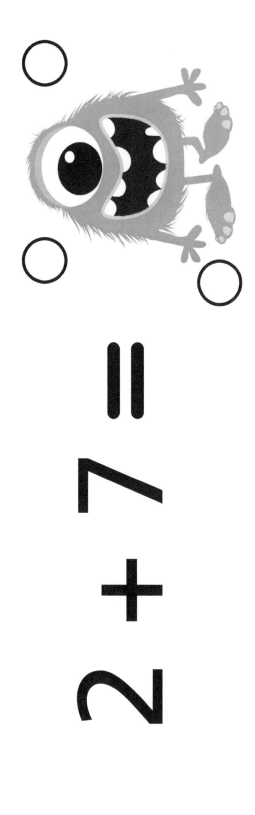

$$2 + 7 =$$

Balance Puzzle 3

$$\bigcirc = 5 - 1$$

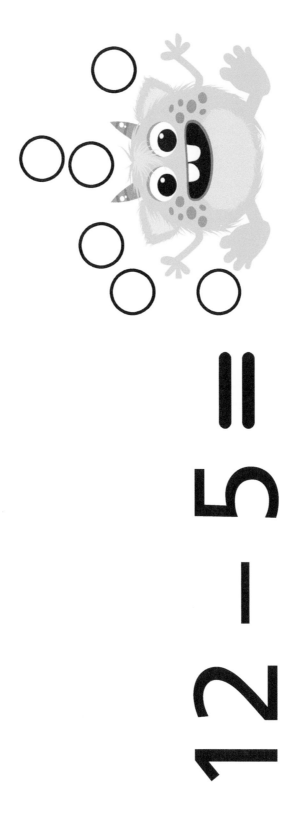

$$12 - 5 =$$

$$7 - 2 =$$

$$= 2 + 14$$

$$12 =$$

= 5 + 9

 **Monster Bank**

True or False?

Snapshot

Students explore the meaning of equivalence by playing True or False? with equations.

Connection to CCSS
1.OA.7, 1.OA.3, 1.OA.5, 1.OA.6

Agenda

Activity	Time	Description/Prompt	Materials
Launch	10–15 min	Do a True or False? number talk with the equation $2 + 3 = 5 + 1$. Give students think time, poll the class for who thinks it is true or false, and then invite students to share reasoning. Support the class in coming to agreement, using evidence and the meaning of the equal sign, that this equation is false. Tell students that they will play with puzzles like this today.	Chart and marker or board for displaying number talk and student evidence
Play	20–30 min	Partners play True or False? by sorting a set of equation cards into the two categories. Students use models and verbal explanations to convince one another whether each equation is true or false. Confer with students to press on reasoning and the use of mathematical language for equivalence.	• True or False? Cards, one set per partnership • Make available: manipulatives for modeling, such as reken-reks, unifix cubes, or chips
Discuss	10–15 min	If students struggled with a particular equation, have them bring it to the class to debate and come to consensus about. You may also want to share compelling models that students created. Discuss the strategies that students developed for deciding whether an equation was true or false and the evidence they found convincing.	Chart and markers

Activity	Time	Description/Prompt	Materials
Extend	30+ min	Partners design their own True or False? card set, using numbers or other visuals to create equations. Partners sort their own cards to check that they have both true and false cards, before trading sets with another group for sorting. This activity can become a center for creating and sorting equations.	• Blank True or False? Cards or index cards, multiple per partnership • Make available: manipulatives for modeling and designing cards, such as reken-reks, snap cubes, chips, dice, or dominoes

To the Teacher

We return to the big idea that equal means the same by playing a game of True or False?, which begins with a number talk. In these puzzles, students are provided with complete equations, some of which are true, in that the two sides are equal in value, and some of which are false, in that the two sides have different values. The students' goal is to determine whether an equation is true or false and provide convincing reasoning for why.

This game often uncovers common misconceptions around the equal sign, particularly that it means "make" and that the number immediately after the equal sign is the answer to what came before. We touched on this possible misconception in the Visualize activity, and if it did not surface there, it is likely to here. For instance, in the launch we begin with a True or False? number talk using the equation $2 + 3 = 5 + 1$. Some students will read across this equation like a sentence, checking that each step remains true, as in $2 + 3$ is equal to 5, true, and then we add 1. Treating the equation as a series of steps undermines the idea of equivalence because students ignore the value of each side of the equation as a whole. Use the number talk to surface this misconception, if present, and allow your students to engage in mathematical disagreement about whether the statement is true or false, rather than stepping in and telling them that it is false. This is the challenging mathematical work of these puzzles, and where students cannot resolve the disagreement, you know it is worthy of continued discussion.

Consider doing this activity repeatedly over a series of days. You could weave back and forth with the Visualize activity, doing balance puzzles and playing True or False? as number talks and partner work. Similarly, these can both become centers for ongoing exploration.

Activity

Launch

Launch the activity by facilitating the following number talk:

$$\text{True or false?}$$
$$2 + 3 = 5 + 1$$

Put this number talk on a chart or board and ask, Is this equation true or false? Read it aloud. Give students time to think and show a silent thumbs-up when they are ready. You may want to poll the class for who thinks it is true and who thinks it is false. Invite students to share their reasoning and evidence. Make connections to the ways that students were thinking about the equal sign and its meaning in the Visualize activity. If the class seems stuck, ask, What does the equal sign mean? If students don't agree, focus on the debate and support students in resolving the disagreement using mathematics. Resist the urge to say who is right and instead ask questions to support students in developing convincing arguments, such as, Is there a way we could model it to see whether it is true or false?

Through the discussion, come to agreement and label this equation false: $2 + 3$ is not equal to $5 + 1$. Ask, What reasoning did you find convincing or helpful? Tell students that their challenge today is to figure out with their partner whether equations are true or false and to focus on giving reasoning to each other that is helpful and convincing.

Play

Provide partners with a set of True or False? Cards and access to tools for modeling, such as rekenreks, snap cubes, chips, or other manipulatives. Partners work together to model, reason about, and sort the equations into true or false categories. Students use the following questions to support their thinking:

- What is the value of each side of the equation?
- How could you model the two sides of the equation to see whether they are equal?

- Is the equation true or false? Why? How do you know?
- What evidence can you share with your partner that will convince them?

As you circulate conferring with students, press them to share their reasoning for either true or false and use mathematical language, such as *equal, same, value,* and *amount,* in their explanations. Encourage students to create models using reken-reks, snap cubes, chips, or other counters to support their thinking and convince each other.

Discuss

Discuss any equations that you noticed, or that students told you, that the class found challenging to decide whether it was true or false. Post the equation where everyone can see it and discuss the following questions:

- Why might this be true?
- Why might this be false?
- How could we prove whether it is true or false?

Discuss until the class can come to agreement as to whether the equation is true or false. Highlight evidence and the use of reasoning about equality that is clear or persuasive.

Alternatively, if you notice that students made interesting or convincing models, you may want to invite them to share these with the class. Highlight what features of the models help us see whether an equation is true or false.

Use the following questions to discuss the strategies students used to sort the equations:

- What strategies did you and your partner use to decide whether an equation was true or false?
- What models did you make that were helpful?
- What evidence did you find convincing?
- What challenges did you face?
- Was there an equation that you changed your mind about? If so, what made you change your mind?

Extend

Partners design their own True or False? Cards for others to use. Provide students with cards from the Blank True or False? Cards sheet or index cards, and access to tools for modeling the equations they create. Students can use numbers, dots, dice, dominoes, or any other visual symbols to design their equations. Partners should be sure to include both true and false cards in their set. Students should sort their own cards before sharing them with others, and when they are confident that they have created both true and false equations, they can trade with another group. This extension can become a center for creating cards and sorting those left by others. Alternatively, you can use the pool of equations students create as number talks for the class.

Look-Fors

- **Are students thinking about the value of the two sides of the equation?** As in the Visualize activity, this game is intended to support students in thinking about the value of the two sides of the equation. In this case, students must ask, Do the two sides have the same value, or not? Students will need lots of opportunities through these two activities to develop this conceptual approach to equations. You can support this developing thinking by asking questions such as, What is the value of this side of the equation? How do you know? What is the value of the other side? What does that mean?

- **Are students using any relational strategies to determine whether an equation is true or false?** One key strategy that this game is meant to support is determining the value of each side of the equation and then comparing those values. For instance, in the number talk in the launch, students can determine that the value of $2 + 3$ is 5 and the value of $5 + 1$ is 6 and then compare 5 and 6 to decide that these are not equal. However, as students gain experience with these kinds of puzzles, you may begin to notice glimmers of another set of strategies, *relational strategies*, which we will return to in the Investigate activity. Relational strategies compare the relationships between the quantities on both sides of the equation, often without calculating the full value of both sides. For instance, students could view the equation in the launch and determine that the value of $2 + 3$ is 5 and then note that on the other side is 5 with something more added to it. We do not need to know its value to determine that it will be greater than, or at least not equal to, 5, the value of the left side. This is an early relational strategy, and it represents a key step toward more algebraic thinking. Be sure to create space in the discussion for students to share these strategies so that others can begin to try them out.

- **What forms of mathematically sound evidence are students providing?**
Students' evidence should be mathematically sound, meaning that it draws on mathematical ideas rather than just being offered by someone other students trust to have the right answer. Students need to provide more than an opinion; they need to offer models, whether they are verbal, physical, written, or drawn, to show what the value of each side of the equation is and why these are equal or not. Students may use fingers, cubes, numbers, words, circles and arrows, rekenreks, or any number of other tools to show others what their thinking is and why it makes sense. When you ask the class whether the evidence makes sense, be sure that this is not a rhetorical question but rather a genuine one to which students might agree, disagree, or offer, "Not yet." Encourage students to ask questions about the evidence students present and remember that even if evidence is convincing to you, it may not yet be persuasive to your class. Invite students to revoice one another's thinking and evidence to hear how they are making sense of one another. If students offer evidence that does not yet make sense to you, feel free to say as much, as in, "I don't know. I'm not convinced yet," and then model how to pose questions about evidence. This kind of mathematical argumentation requires a good deal of practice across many years to develop, and this is a good place to start.

Reflect

How do you know whether an equation is true or false?

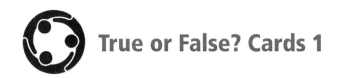

True or False? $8 = 10 - 2$	True or False? $6 - 3 = 1 + 2$
True or False? $7 + 2 = 4 + 5$	True or False? $9 - 3 = 6 - 2$
True or False? $4 + 6 = 10 - 4$	True or False? $7 + 1 = 1 + 7$

True or False?

$$2 + 6 = 11 - 3$$

True or False?

$$0 + 6 = 6 - 0$$

True or False?

$$3 + 3 = 4 + 3$$

True or False?

$$9 - 5 = 3 + 2$$

True or False?

$$7 = 9 - 3$$

True or False?

$$10 = 2 + 6$$

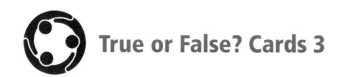

True or False? Cards 3

True or False? $5 - 3 = 6 - 2$	True or False? $6 + 4 = 7 + 2$
True or False? $10 - 2 = 11 - 3$	True or False? $6 + 3 = 4 + 5$
True or False? $5 - 4 = 4 - 3$	True or False? $12 + 3 = 11 + 3$

Blank True or False? Cards

True or False?	True or False?
True or False?	True or False?
True or False?	True or False?

Exploring Relations

Snapshot

Students investigate finding the missing value in equations to develop relational reasoning.

Connection to CCSS
1.OA.7, 1. OA, 8, 1.OA.3, 1.OA.4

Agenda

Activity	Time	Description/Prompt	Materials
Launch	10 min	Do the following number talk, in which students try to find the missing number: 12 + 4 = __ + 5. Take students' answers and reasoning and highlight any examples of relational reasoning. If students do not share any, ask, Is there any way we could have figured out the missing number without finding the value of each side? Discuss possibilities.	Optional: chart and markers
Explore	20–30 min	Provide partners with a set of Missing Number Cards and access to tools for modeling. For each equation, students try to find the missing number. Students investigate how to find the missing number without finding the value of each side of the equation. Partners can then try to make their own equations that do not require finding the value of each side to solve.	• Missing Number Cards, set of 4–8 per partnership • Make available: manipulatives for modeling, such as snap cubes, rekenreks, and chips • Optional: Make Your Own Missing Number Cards or index cards
Discuss	10–15 min	Discuss the ways that students thought through solving the missing-number problems, highlighting any strategies that use relational thinking.	

To the Teacher

In this investigation, we open the door to relational thinking with the equal sign. Up to this point, the activities in this big idea have focused on developing the concept of the equal sign so that students understand that the value of both sides must be the same. Evaluating this, as students did in the Play activity, typically involves calculating the value of each side and comparing, or using the value of one side to find missing numbers on the other, as in the Visualize activity. This is precisely the work we want students to be learning to do. Once they have learned it, however, they may be ready to begin to develop relational thinking.

Relational thinking involves using the relationships between the quantities on both sides of the equal sign and knowledge of properties to solve problems. For instance, to solve a problem like $24 + 37 = 37 + \_\_$, you could find the sum on the left and then use that to find the missing number on the right, but it is not necessary. Instead, thinking relationally involves noticing that order does not matter when adding (commutative property), so adding 24 to 37 must be the same as adding 37 to 24. In this case, we never have to know the sum of $24 + 37$ to find the missing number. Another way we use relational thinking is when values are similar, as in $19 + 6 = 18 + \_\_$. In this example, instead of finding the sum, we might notice that 18 is just 1 less than 19. That would mean that the missing value must be 1 greater than 6; it is as though we shifted 1 from the 19 to the 6, creating $18 + 7$, which has the same value. You can see both how powerful this thinking is and what an advancement in reasoning about number and operations it represents. If you can think relationally, you can solve problems far more complex than those you can compute. For instance, a student could find the missing value in the following equation, even if they did not yet know how to subtract quantities this large: $101 - \_\_ = 100 - 56$. If you'd like to learn more about relational thinking, we recommend reading *Thinking Mathematically: Integrating Arithmetic and Algebra in Elementary School* (Carpenter, Franke, & Levi, 2003).

Note that we encourage you to keep both avenues for reasoning about the equal sign open in this activity. Students can find and compare the values of both sides or reason relationally. The goal is not to take away one of these tools and substitute the other. Rather we want students to be exposed to the idea that, sometimes, there are ways of solving problems of equality that do not require calculating. This is a big developmental leap, and not all students will make it at the same time. Instead, continue to engage students in thinking in a variety of ways about equality and to ask the kinds of questions that invite relational thinking.

Activity
Launch

Launch the activity by posting the following number talk on a chart or board:

$$12 + 4 = \underline{} + 5$$

Ask, What is the missing number to make the equation true? Give students time to think and show a private thumbs-up when they are ready. Invite students to share their answers, but the key here is the reasoning students have used. Students can solve this by finding the sum of $12 + 4$ and then using that sum to find the missing value. This is accurate and should be recognized as such, but it is not relational thinking. Highlight any relational thinking students do, in which they consider the relationship between 5 and the numbers on the left side of the equation. For instance, relational reasoning here would sound something like, "Five is 1 more than 4, so the number added to it will need to be 1 less than 12."

If you do not hear any relational thinking offered, then you might ask, Is there any way we could have figured out the missing number without finding the value of each side? Students may need time to think or turn and talk to a partner to discuss possibilities. Be sure in the discussion of the number talk not to devalue calculation strategies, but rather to encourage students to think of other ways for solving too.

Explore

Provide partners with a set of Missing Number Cards and access to manipulatives for modeling, such as snap cubes, rekenreks, and chips. Partners investigate each equation using the following questions:

- How can you find the missing number to make the sentence true?
- What different strategies could you use to find the value that is missing?
- Is there a way to find the missing value without finding the value of each side of the equation?

After students have had time to explore these equations, you might have partners try creating their own: What missing-number equations can you write that you can solve without finding the value of each side of the equation?

As a class, discuss the following questions:

- How did you find the missing numbers? What strategies did you develop?
- How did you use the equal sign to help you reason about what was missing?
- Did anyone find the missing number without finding the value of each side of the equation? How?

Highlight examples of relational reasoning, in which students thought about the relationship between the quantities on each side without necessarily finding the sums or differences on either side.

Look-Fors

- **Where do you notice students beginning to use relational thinking?** Students may first begin to use relational thinking in missing-number problems that involve the same numbers, such as $15 + 16 = 15 + \_\_$ or $19 + \_\_ = 13 + 19$. Alternatively, some students may begin to reason relationally with smaller numbers that are most familiar, as with $8 + 5 = 7 + \_\_$. Still other students may be prompted to use relational thinking when the numbers involved in finding the value of one side seem daunting and they are seeking a more efficient route, as with $12 + \_\_ = 11 + 28$. Remember that developing relational thinking is a process, and students may use relational thinking with particular problems and not with others. Be sure to probe student reasoning where you see relational thinking to support students in articulating why their solution makes sense and what they noticed about the problem that helped them think relationally.

- **Are students using models to support relational thinking?** While students may begin to think relationally, they may still want to use manipulatives to confirm that their reasoning makes sense or to communicate their thinking to others. Relational reasoning is a valuable tool, and it will be even more useful if students can visualize the relationships they are using to solve the problem. If students struggle to explain their relational reasoning, ask them to show it with tools. For instance, to solve $12 + \_\_ = 11 + 28$, students might have imagined 1 (cube, chip, or unit) moving from the 28 to the 11, transforming the expression into $12 + 27$. Students can show this movement with manipulatives or through drawings, with or without counting out all 11 and 28 objects. Here it is the movement or transfer that matters, and students may be able to explain this idea through objects or images. Alternatively, students may find a missing number using relational reasoning, but not yet trust this process and therefore need to see that it works by counting out snap cubes to confirm the answer. In all cases, we encourage you to support students in making connections between numbers, objects, operations, and relationships, layering on different representations, rather than taking any away.

Reflect

When can you solve a missing-number problem without finding the value of each side of the equation? Give an example if it helps you explain.

Reference

Carpenter, T. P., Franke, M. L., & Levi, L. (2003). *Thinking mathematically: Integrating arithmetic and algebra in elementary school*. Portsmouth, NH: Heinemann.

15 + 16 = 15 + ☐

19 + ☐ = 13 + 19

8 + 5 = 7 + ☐

12 + ☐ = 11 + 28

13 + 5 = 12 + ☐

10 + 7 = 9 + ☐

$\boxed{} - 17 = 15 - 15$

$11 - \boxed{} = 10 - 3$

$10 - \boxed{} = 21 - 3$

$\boxed{} - 2 = 33 - 3$

$7 + 7 = 8 + \boxed{}$

$\boxed{} + 16 = 7 + 15$

$14 + 22 = \boxed{} + 32$

$18 + 8 = 38 - \boxed{}$

$33 - 5 = 34 - \boxed{}$

$29 - 6 = \boxed{} - 7$

$25 - \boxed{} = 26 - 6$

$32 - 3 = \boxed{} - 4$

Make Your Own Missing Number Cards

Building with Numbers within 20

Venod Menon (2015) is a Stanford neuroscientist who has conducted important research with his team, including Lang Chen, a neuroscientist I have collaborated and authored papers with. One area of their fascinating work shows that our brain has six different regions at work when we think about mathematics. Two of the regions specialize in visual thinking. It is essential to engage all of these brain areas well when students work on a mathematics task, which is one reason we highlight visual mathematical thinking in our tasks in these books and on youcubed. org. Something else that their work has shown to be important is opportunities to connect different brain regions. This happens when we encounter mathematics in different ways—for example, when we see a number as a symbol and in visuals. Researchers Joonkoo Park and Elizabeth Brannon (2013) found that different areas of the brain were involved when people worked with symbols, such as numerals, than when they worked with visual and spatial information, such as an array of dots. The researchers also found that mathematics learning was optimized when these two areas of the brain were communicating with each other. We can learn mathematical ideas through numbers, but we can also learn them through words, visuals, models, algorithms, tables, and graphs; from moving and touching; and from other representations. But when we learn by using two or more of these means and the different areas of the brain responsible for each communicate with one another, the learning experience is greatest.

In this big idea, we give students different ways to experience numbers, moving now to double-digit numbers. In Big Idea 5, students were developing ways to model and understand the equal sign; now they get to experience double-digit numbers, which should be very exciting!

In our Visualize activity, we introduce one of my favorite mathematical activities—a dot-card number talk. I use these with students of all ages, even my Stanford undergraduates. If you would like to see me teach a class of middle school girls with a dot-card number talk, you can find the video here: https://www.youcubed.org/resources/jo-teaching-visual-dot-card-number-talk/. This is a perfect activity for promoting brain connections, as students see numbers and their visual equivalents. After the number talk, students see groups of the same dot pattern and work to color-code it to show different ways of seeing it. The discussion should include doubles, such as 5 + 5 or 6 + 6, pairs that make friendly numbers, and all sorts of fun number flexibility.

The Play activity is a game that students really enjoy and that will give them meaningful opportunities for conceptual understanding. This is a dice game inspired by the game called Tenzi. Students roll dice and record their number into a 5 × 5 number grid where they look for ways to make 10. They color-code and highlight the 10s as they try to fill the grid and combine all of the numbers entered. This is a game where the lowest score wins because students record the number of dice-roll entries they could not use to make a 10.

The Investigate activity is where we introduce the addition table. Students will be asked to look for patterns, color-code the table, and describe their patterns. We also share an extended table beyond 10 so that students can explore numbers that will take them to sums beyond 20. The goal is for students to recognize patterns in the table, model with manipulatives the patterns they see, and share their findings. Teachers can encourage students to push their thinking by asking them to think about the patterns they find and consider whether their patterns would continue.

Jo Boaler

References

Menon, V. (2015). Arithmetic in the child and adult brain. In R. C. Kadosh & A. Dowker (Eds.), *The Oxford handbook of numerical cognition* (pp. 502–530). Oxford, UK: Oxford University Press.

Park, J., & Brannon, E. M. (2013). Training the approximate number system improves math proficiency. *Psychological Science, 24*(10), 2013–2019.

Seeing Spots

Snapshot

Students explore decomposing numbers between 10 and 20 using dot images, color-coding and labeling them to show multiple ways to compose these numbers.

> **Connection to CCSS**
> 1.OA.5, 1.OA.6, 1.NBT.2b

Agenda

Activity	Time	Description/Prompt	Materials
Launch	10 min	Facilitate a dot talk with the image of 10 dots. Ask students how many dots they see and how they see the dots. Color-code each way of seeing the dots that students share and post these side by side. Ask students how they could represent one of these ways of seeing the dots with a number sentence and label the dot sheets accordingly. Draw attention to equivalence.	• Dot Talk sheet, multiple copies • Markers and tape
Explore	20+ min	Partners explore the three different ___ Dots sheets to determine how many dots the images show and how they see the dots. Students color-code the representations of the dot arrangements to show how they saw the number of dots and label each with a number sentence.	• ___ Dots sheets, each of the three sheets available, per partnership • Colors, per partnership
Discuss	10–15 min	Invite students to share the different ways they saw the same images of dots. Record these different ways on a class ___ Dots sheet, including the corresponding number sentences. Discuss whether these number sentences are equal.	• ___ Dots sheets, to display • Colors

To the Teacher

Dot talks are a powerful routine for developing subitizing, decomposition, and composition of number. Students often make connections between shapes and numbers, such as triangles and the number 3, when looking at the arrangement of dots and asking, How many? These connections support students in thinking visually about number and maintain the connection between the physical objects and the number that we use to represent them. There are many guides to dot talks, such as Parrish's (2010) *Number Talks*, and you can easily create your own.

In this activity, we begin with a number talk focused on 10 dots arranged in a way that encourages several possible decompositions of the number 10. Students may use the symmetry of the arrangement in different ways, draw on their knowledge of dice to see 4 or 6 in the center, or decompose in rows, among many other strategies. Each of these ways can lead to a different number sentence to represent the particular way that students decomposed the image to then recompose 10. In the launch, your work to surface, highlight, and label these different ways can help students see that there are many ways to make 10.

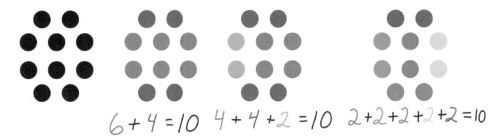

Three different ways of seeing the 10-dot pattern

For the exploration, we have constructed three different types of dot arrangements for numbers between 10 and 20: 12 (Dots Sheet 1), 15 (Dots Sheet 2), and 18 (Dots Sheet 3). Each of these numbers is a useful one to be familiar with and builds on students' understanding of 10. For instance, it is useful to understand 15 both as 10 and 5 and as three 5s. Instead of giving each of these sheets a title with its number of dots, we have left the titles with a blank that students might complete as they figure out how many dots are in each image. The sheets each include multiple representations of each arrangement, so that students can decompose the images in multiple ways, each time color-coding and labeling them differently. It may make the discussion more coherent if you begin by providing all students the same image and then offer the other two as choices after students have tried the first.

You can extend this activity by continuing with dot talks that connect counting with number sentences or by making your own sheets with different numbers of dots or different arrangements of the same numbers.

Activity

Launch

Launch the activity by facilitating the dot talk with 10 open dots. We recommend that you have multiple copies of the Dot Talk sheet, markers, and tape available, so that you can color-code the sheets in multiple ways and hang them up to compare side by side. Tell students that you are going to show them an image of dots and that you want to know, How many dots do you see? Show students the Dot Talk sheet for a few moments, then put the sheet down and ask students, How many dots did you see? Take some student answers and then post the sheet for all to see. Ask, How did you see the dots? Invite students to describe or come up and show the different clusters they saw. Color-code the different groups of dots they saw to help others see them the same way.

Point to the color-coded sheets and ask, How could we represent these ways of seeing the dots with number sentences? Give students a chance to turn and talk to a partner. Take some student suggestions and label at least one of the images with one or more number sentences. If students name different but equivalent number sentences (such as $2 + 3 + 5$ and $5 + 2 + 3$), be sure to record them both and use an equal sign to show their equivalence.

Explore

Provide partners with one of the ___ Dots sheets. Students explore the following questions:

- How many dots are there?
- How do you see the number of dots?
- How can you color-code the dots to show how you see them?
- What equation could you write to represent how you saw the dots?

Partners try to find as many ways to see the dots on this image as possible, color-coding and labeling each. Students can then try other dot arrangements. Make available the other ___ Dots sheets for students to explore.

Discuss

Using the following questions, discuss the different ways that students saw the dots on the ___ Dots sheets:

- How many dots are there?
- How did you see the number of dots?
- How did you color-code the dots to show how you saw them?
- What equation did you write to represent how you saw the dots?
- How did you know that your equation matched the way you saw the dots?

Invite students to show their representations on the document camera. You may want to keep a class record of the different ways by recording the different representations side by side on a class ___ Dots sheet.

Looking at the multiple number sentences for the same dot sheet, discuss these questions: Are these equal? Why or why not? How do we know? Use this discussion to return to the big idea of equivalence, supporting students in seeing that if $6 + 6 = 12$ and $2 + 4 + 2 + 4 = 12$, for instance, then $6 + 6 = 2 + 4 + 2 + 4$.

Look-Fors

- **Are students' counts of the dots consistent across the same images?** On any given sheet, students will try to see the number of dots in multiple ways, but these ways should result in the same total count of dots. That is, on the sheet with 12 dots, each image is the same and should always be counted, regardless of the way students count, as a total of 12 dots. One of the underlying ideas in this activity, and in all dot talks, is to promote the idea of equivalence. The dot arrangement is equivalent to the number 12, for instance, which is also equivalent to $4 + 4 + 4$ or $6 + 6$, and so on. These ideas get disrupted if students sometimes count 12 dots and other times count 11 or 13. As you circulate, look for consistency across the total counts and ask questions that highlight equivalence. Some students might label each image with an expression, such as $3 + 3 + 3 + 3$, and in these instances, it makes sense to follow up by asking, How many dots are there in total? These questions close the loop on equivalence and allow you to informally assess whether students are counting consistently.

- **Are students double-counting any dots?** When students focus on the shapes made by the dots, it can be easy to notice that different shapes overlap. For instance, in the number talk used in the launch, students might see two triangles on the top of the image. However, these two triangles share a vertex in the middle, and if students count each one as three dots, they could conclude that there are six dots on the top half of the image instead of five. Color coding can support students in attending to which dots they have already counted, so that they do not double-count any dots. Alternatively, students can double-count dots, but need to then think about how to compensate for this double count. For instance, with the two triangles from the number talk, students could see this as two 3s, but know they need to take away 1 for the dot they have counted twice. This could be represented as $3 + 3 - 1$, which is a mathematically accurate and sophisticated way of thinking about compensation. In this sense, students don't need to avoid double-counting; they just need strategies to be aware of and compensate for it. You might ask questions to point out any instances of double-counting and help students generate ways of compensating, such as, It looks like you counted this dot twice. Is that right? What does that mean? What could you do about that?

- **Do students' number sentences match the ways they have decomposed the number of dots?** As you circulate, be looking for connections between the ways that students decomposed the dot arrangements visually, how they color-coded the image, and the number sentence they used to label it. These three ways of counting and representing should be aligned. But you may find instances where students disconnected these by color-coding their visual interpretation of the dots and then simply writing any number sentence for the total. For instance, student may see the 12 dots as four groups of three dots, but then write a number sentence like $6 + 6$ that is not representative of their visual approach. Ask questions to draw attention to the connections you want students to make, such as, How does this number sentence match the way you saw the dots? How did you see the dots? What groups did you see first? Next? How did you put those groups together? How can you show that with a number sentence?

Reflect

When you see a group of dots, what do you notice first? What kinds of groups are most helpful for counting? Why?

Reference

Parrish, S. (2010). *Number talks: Helping children build mental math and computation strategies, grades K–5*. Sausalito, CA: Math Solutions.

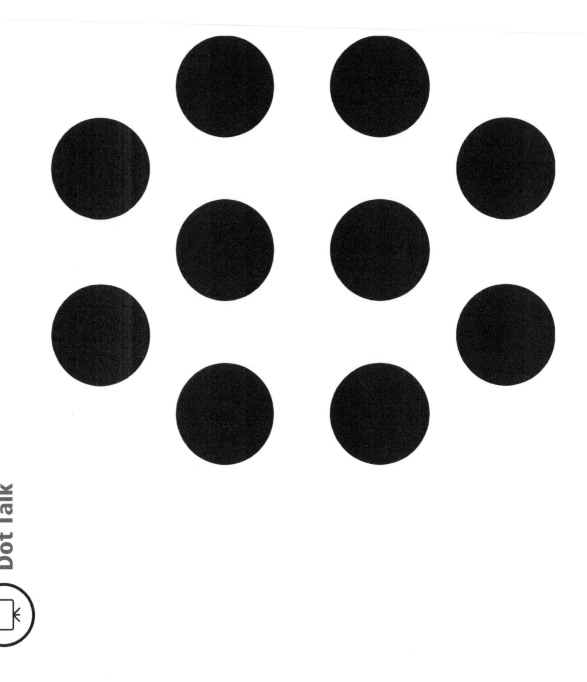

__ Dots: Dots Sheet 1

Dots: Dots Sheet 3

Rolling Ten

Snapshot

Students build fluency with composing and decomposing 10 by playing the dice game Rolling Ten.

Connection to CCSS
1.OA.5, 1.OA.6, 1.OA.3, 1.OA.4

Agenda

Activity	Time	Description/Prompt	Materials
Launch	10–15 min	Show students how to play Rolling Ten using the Rolling Ten Rules sheet. Play the game as a class for a few rounds to ensure that students understand how they might think through placing a number, making 10, and shading in the squares.	• Rolling Ten Rules sheet, to display • Class Rolling Ten Game Board, to display • Highlighters or markers • Die
Play	20+ min	Partners play Rolling Ten collaboratively, shading in groups of numbers that make 10 as they go. Partners play the game repeatedly, trying to fill the board with 10s and get the lowest score possible.	• Rolling Ten Game Boards, multiple sheets per partnership • Colors, per partnership • Die, per partnership • Make available: counting manipulatives, such as snap cubes and rekenreks
Discuss	10 min	Discuss the strategies that students developed for placing numbers on the game board to make 10. Pay particular attention to how strategies may change from the beginning to the end of a game, when fewer squares are available.	

Activity	Time	Description/Prompt	Materials
Extend	45+ min	Students design a variation on Rolling Ten by modifying the rules, target number, game board size or composition, or the number of dice used. Partners test their own games, making revisions until they find their new version interesting and enjoyable. Discuss what kinds of rule changes made the new games fun, and not.	Make available: Blank Rolling Ten Game Boards, grid paper (see appendix), colors, dice, and counting manipulatives

To the Teacher

Building fluency with composing and decomposing numbers begins with making 10 in multiple, flexible ways. Ten is the basis of our counting system, and students can both make 10 and use 10 to make larger numbers. Being comfortable and familiar with the numbers that make a 10 will support students as they move to composing with larger numbers. In this game, students use a die to roll numbers that they can use in different ways to make 10. They need to think strategically about how to place those numbers on a grid to either make 10 immediately or to leave open the possibility of making 10 with a future roll.

We've made this game cooperative, so that students work together to try to make 10 in as many ways as possible, covering the board with their 10s. This enables partners to talk to each other about the different places they could record a number and the advantages of each placement. Each time they think through the possibilities, students will be grappling with composing numbers and thinking about how they relate to 10. For instance, in deciding where to place a rolled 4, students might ask themselves, What pairs with 4 to make 10? They might also ask, Where do I see 6 or numbers that make 6 on the board? If I cannot make 10 now, what number could I position 4 near to be able to make 10 with a future roll? If I place it next to a 5, how many more do I need to make 10? What if I place it next to a 3; how many more do I need to make 10 then? The sheer volume of thinking involved can lead to interesting conversations, the development of strategies, and lots of flexible thinking about number relationships.

Activity

Launch

Launch the activity by modeling the rules of the Rolling Ten game. Show students the Class Rolling Ten Game Board and tell students that their goal is to make 10 using squares that touch each other along a side. Partners will roll a die and then record the number in any empty square. When they make 10, they color in the squares that make that 10. Show students what kind of arrangements count as making 10 and be clear that diagonals do not count as touching along a side. Point out examples of this on the Rolling Ten Rules sheet, and make sure students see that the two 5s that touch on a corner only are not colored in as a 10.

Play the game as a class on the Class Rolling Ten Game Board. Roll the die and ask the class where the number could be placed and why. Be sure to explore multiple options before deciding where to place the number. With some rolls, the placement may be clear, because a 10 can be made; with other rolls, there may be several equally useful placements. Play the game until the class makes at least one 10 and then show students how to shade. Tell students that they should shade each group of 10 a different color so that they can all see all the different ways they made 10.

Play

Provide partners with Rolling Ten Game Boards, a die, colors, and access to counting manipulatives such as rekenreks or unifix cubes. Partners play Rolling Ten collaboratively.

Rolling Ten Rules

- Roll the die and place the number anywhere on the game board, with the goal of making adjacent squares add to 10.
- When you can build 10, shade in the squares that make 10. Try to make each group of 10 a different color.
- The game is over when all the squares are filled and you cannot make any more 10s.
- Your score is the number of unshaded squares at the end. Record your score next to the game board. What is the lowest score you and your partner can get?

As you confer with students, point to an empty square and ask, What number would you like to put in this square to make 10? How do you know? Encourage students to model these situations with cubes or on the rekenrek to strategize or determine what numbers they need to roll to make 10.

Discuss

After students have had a chance to play several rounds of the game, gather as a class to discuss the following questions:

- What strategies did you come up with for making 10?
- How did you decide where to put numbers?
- What happened toward the end of each game? How did your decisions change? How did you know where to place the final numbers?
- Did your scores get better as you played? Why or why not? What did you learn?
- What ways to make 10 did you find? (Record as many different ways that students found as you can on a class chart.) Are these equal? Why or why not?

Extend

Invite students to invent a variation on this game. They might do any of the following:

- Change the size of the board.
- Change the numbers that are placed in the board in advance.
- Roll with more than one die.
- Change the target number from 10 to something else, such as 12 or 20.
- Change the rules. For instance, students might decide that if you roll two dice, the numbers must be placed side by side.

Provide students with Blank Rolling Ten Game Boards, grid paper (see appendix), dice, colors, and counting tools to experiment with and develop their own games. Partners work together to design a variation and test it themselves. Be sure to allow students to iterate their designs, learning from what does and does not work well to design a game. Students can play their own games, share them with the class, and try others' games. Discuss what rules they decided were most interesting and why.

Look-Fors

- **Are students placing numbers strategically?** Students need to be thinking about where they could place the numbers they roll to make 10, rather than locating them at random. At times, a random placement may make the most sense, particularly in the beginning of the game if the number rolled does not help make 10 with any existing clusters of numbers. However, even this decision can be intentional; students might determine that the number they rolled doesn't help them in any specific way and might be best placed on its own. As you circulate, watch students roll, and listen to the ensuing conversation. Are partners discussing different possible placements? What ideas are they considering? What are they looking for on the board? Do you hear students reject placements or propose others that are better in some way? Ask students about their reasoning at any stage to uncover their developing strategies. You might ask, Why did you put this number here? Where could you put the number you just rolled? What are you looking for? Why?

- **Are students thinking about the numbers they need to roll to make 10?** Toward the end of the game in particular, as the board gets crowded and only a few spaces remain, students may be thinking about which numbers they want to roll to fit into those spots to make a 10. When you encounter a partnership near the end of the game, pause them to ask, What number do you want to roll to fit in this spot to make 10? How do you know? You might ask students to consider each of the remaining spots to build a short list of hoped-for rolls before they roll the die. Some students may do this independently, while others may be encouraged by your questions.

- **Are students identifying frequent combinations for making 10?** As students play, one goal is that they begin to become fluent with some commonly found combinations of 10. With only one die, the combinations are somewhat restricted when combining two numbers; only 5 + 5 and 6 + 4 are possible. Learning these two combinations through the game will help students play and is a natural outcome of play. You do not need, nor should you try, to teach students to memorize these in advance. Rather, students will become fluent precisely because these are useful. Similarly, you may see students becoming fluent with some other combinations of three or four numbers, such as 5 + 3 + 2 or 3 + 3 + 3 + 1. When asked why they want to place a number in a particular spot, students may share, over time, that they just know that these numbers go together to make 10. This is how fluency develops naturally.

Reflect

Which ways of making 10 were most useful or common when you played Rolling Ten? Why do you think that is?

1. Roll 1 die.

2. Write the number in a box.

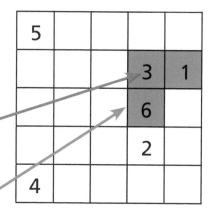

3. Color numbers that make 10. The numbers must share a side.

A number can only be used once to make 10.

4. Game ends when all boxes are full.

5. Low score wins!

Score = 6

Scoring: Count the number of boxes that are not colored. That's your score!

Class Rolling Ten Game Board

	5		3	
1			4	

Rolling Ten Game Boards

<table>
<tr><td></td><td></td><td></td><td></td><td></td></tr>
<tr><td></td><td></td><td>3</td><td>5</td><td></td></tr>
<tr><td></td><td></td><td></td><td></td><td>4</td></tr>
<tr><td></td><td></td><td></td><td></td><td></td></tr>
<tr><td></td><td></td><td>6</td><td></td><td></td></tr>
</table>

<table>
<tr><td></td><td>4</td><td></td><td>5</td><td></td></tr>
<tr><td></td><td></td><td></td><td></td><td>1</td></tr>
<tr><td></td><td></td><td>6</td><td></td><td></td></tr>
<tr><td></td><td></td><td></td><td></td><td></td></tr>
<tr><td></td><td></td><td></td><td></td><td></td></tr>
</table>

<table>
<tr><td></td><td></td><td></td><td></td><td></td></tr>
<tr><td></td><td></td><td>2</td><td></td><td>4</td></tr>
<tr><td>5</td><td></td><td></td><td></td><td></td></tr>
<tr><td></td><td></td><td></td><td></td><td>5</td></tr>
<tr><td></td><td></td><td></td><td></td><td>1</td></tr>
</table>

<table>
<tr><td></td><td>6</td><td></td><td></td><td>2</td></tr>
<tr><td></td><td></td><td></td><td>5</td><td></td></tr>
<tr><td></td><td></td><td></td><td></td><td></td></tr>
<tr><td></td><td>4</td><td></td><td></td><td>3</td></tr>
<tr><td></td><td></td><td></td><td></td><td></td></tr>
</table>

Blank Rolling Ten Game Boards

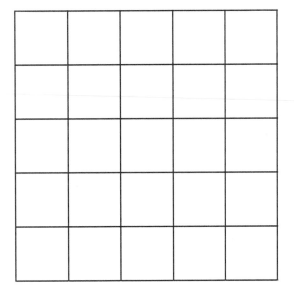 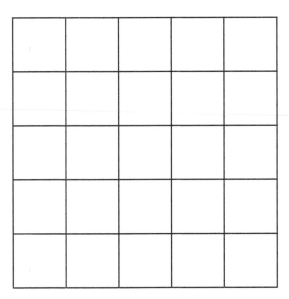

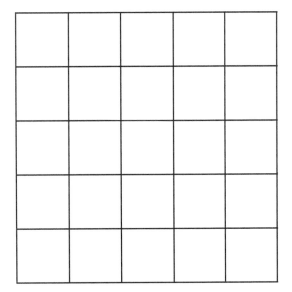 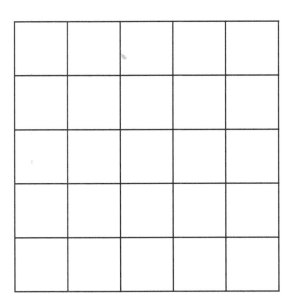

Addition Table Patterns

Snapshot

Students investigate patterns in the addition table, building algebraic thinking, an understanding of the properties of addition, and flexibility with joining quantities.

> **Connection to CCSS**
> 1.OA.6, 1.OA.3, 1.OA.5

Agenda

Activity	Time	Description/Prompt	Materials
Launch	10 min	Show students the Addition Table sheet on the document camera and engage students in making sense of what the table shows and how it is organized. Be curious together about the information it holds and what it can tell us.	Addition Table sheet, to display
Explore	25–30 min	Using an Addition Table sheet, partners explore, color-code, and model the patterns they see. Partners use manipulatives to explore what might lead to these patterns. Invite partners to predict and test whether these patterns would continue if the table were extended.	• Addition Table sheet, per partnership • Colors, per partnership • Make available: counting manipulatives such as square tiles, reken-reks, and snap cubes • Extended Addition Table sheet, per partnership, as needed
Discuss	10–15 min	Discuss the patterns students noticed, what generates those patterns, how we can describe them, and whether they might continue if the table were extended. Encourage students to use tools to show why these patterns exist.	• Addition Table sheet, to display • Colors • Make available: counting manipulatives such as square tiles, reken-reks, and snap cubes

Activity	Time	Description/Prompt	Materials
Extend	30+ min	If during the discussion students developed interest in exploring additional questions, create space and opportunity for them to continue the investigation.	Make available: counting manipulatives such as square tiles, rekenreks, and snap cubes; colors; grid paper (see appendix); or other tools, as needed

To the Teacher

The patterns we see in the addition table seem straightforward to us as adults, but they are novel to young children, and they may never have seen addition represented in this way. Indeed, tables themselves, as tools for organizing data, are still relatively new, and one this size may be entirely new. When we ask students to look at such a complex representation of 121 joining combinations in one image, students may both see an enormous variety of patterns and make individual, specific observations. Embrace whatever students justify, even if unconventional. Press students to articulate their ideas and then invite students to explore those ideas, whether or not they are conventionally stated or commonly held. Encourage a spirit of curiosity and exploration.

This activity foreshadows a similar exploration of the multiplication table in fifth grade. Looking for patterns in tables is one that will support students in the long run as they develop ideas about data, patterns, algebraic thinking, and number relationships.

Activity

Launch

Launch the activity by showing students the Addition Table sheet on the document camera and asking them to interpret it. Ask, What do you think this shows? What do you notice? Invite students to turn and talk to a partner about what they see. Ask students to share ideas, whether they are specific observations or overall interpretations. Come to a shared understanding of this image as a table that shows the sums of numbers. Interpreting this as a table is difficult, so spend time on this, including what the rows, columns, and intersection points mean. While we want students to have a clear understanding of what the Addition Table sheet shows, it is just as important that students make observations and ask questions. Cultivate a sense of curiosity about this representation.

Explore

Provide partners with the Addition Table sheet, colors, and counting manipulatives such as square tiles, rekenreks, and unifix cubes. Students explore the following questions:

- What patterns do you see?
- What makes these patterns?
- How can you show these patterns using snap cubes, rekenreks, or square tiles?

Partners explore, color-code, or mark up the Addition Table sheet with the different patterns they find. Students may need more than one sheet to track the various patterns they notice, so that their markings do not become too dense or jumbled. Students can use manipulatives to investigate what the table shows and where the patterns come from.

Invite students to make predictions about their patterns, by exploring the question, Would these patterns continue if you extended the table? Provide partners with the Extended Addition Table sheet so that students can explore, extend, and test ideas.

Discuss

Gather students together to discuss the following questions, using an Addition Table sheet on display for marking or reference:

- What patterns did you notice? (Invite students to show those patterns on the table and describe them with words.)
- Why do these patterns exist? How did you show them with manipulatives?
- Will these patterns continue forever? Why or why not?
- How did you test your ideas? What did you find?
- What questions do you have now about the addition table? How could you explore those questions?

During this discussion, press students to use increasingly precise language for what they notice. For instance, students might say that the numbers "go up"; while you talk, press students to say more about what it means to "go up" and by how much. You may, as a class, be able to find more precise ways of describing this pattern, such as, "As you go down the column, the sum goes up by 1." Similarly, it is critical that students make sense of why these patterns exist. For instance, why is it that when you go down the column, the sum increases by 1? What is happening? Encourage students to show what is happening using tools to help everyone make sense of the patterns.

Extend

If students developed interesting predictions or questions during the exploration or discussion, offer them an opportunity to continue to explore these ideas. They may need space to extend the table, such as sheets of grid paper (see appendix), or they may need manipulatives for modeling the patterns. For instance, some students may want to explore what new patterns might arise if the table included numbers up to 20 + 20; others may ask, What could a subtraction table look like? Help students articulate the questions they want to ask and provide the tools they need to continue the investigation.

Look-Fors

- **Are students interpreting the features of the table accurately?** Although we do not want to lead students to see or prioritize particular patterns, we do need to ensure that they understand what the table shows. Students need to understand that the rows and columns represent quantities that are being joined, and that the sum of each pair of numbers can be found where the row and column intersect. This is not intuitive for young children who have limited experience with tables, or whose experience may be confined to T-charts. We encourage you to use the launch to address how to interpret the table, but during the exploration, you may encounter students who are struggling, not with patterns, but with understanding the table as a tool. Slow down and support students in making sense of the table. Students may benefit from focusing on the meaning of individual squares before seeing the overall meaning of the table.

- **Are students noticing patterns along with making individual observations?** Both individual observations, such as 3 + 4 = 7, and observations about patterns, such as that the row on the top is the same as the column on the left, are important sites for making sense. Individual observations support students in making sense of what the table shows and what it means, and patterns help students think about the relationships within the table and about addition more generally. Students won't be able to make sense of the patterns if they cannot yet make sense of the individual pieces of data they contain. As you circulate, be looking for both individual observations and observations about overall patterns, and how they support each other. Ask students questions about the individual pieces of data inside the patterns they see, and about how the individual observations accumulate into patterns.

- **Are students figuring out how to model patterns with manipulatives?** One way that students can make sense of what the patterns they see mean in the world is to model them. Students might, for instance, use a rekenrek to show 1 + 1, and then transform that into 1 + 2, 1 + 3, 1 + 4, and so on, to show the pattern that occurs in the top row of the addition table. Students could model this same pattern with snap cubes, with two different colors used to show the two addends, making a stick of one blue and one red, then another with one blue and two red, and so on, arranged side by side to show the pattern of growing. In each case, manipulatives can physically model the patterns that students observe in the table, providing evidence for why they exist and whether or not they might continue. Ask students, How could you show this pattern with objects? How could you show the first part of the pattern (e.g., 1 + 1)? And the next part? How could showing them help you understand what this pattern means?

Reflect

What are you wondering about the addition table now? How could you investigate?

Addition Table

+	0	1	2	3	4	5	6	7	8	9	10
0	0	1	2	3	4	5	6	7	8	9	10
1	1	2	3	4	5	6	7	8	9	10	11
2	2	3	4	5	6	7	8	9	10	11	12
3	3	4	5	6	7	8	9	10	11	12	13
4	4	5	6	7	8	9	10	11	12	13	14
5	5	6	7	8	9	10	11	12	13	14	15
6	6	7	8	9	10	11	12	13	14	15	16
7	7	8	9	10	11	12	13	14	15	16	17
8	8	9	10	11	12	13	14	15	16	17	18
9	9	10	11	12	13	14	15	16	17	18	19
10	10	11	12	13	14	15	16	17	18	19	20

Extended Addition Table

+	0	1	2	3	4	5	6	7	8	9	10				
0	0	1	2	3	4	5	6	7	8	9	10				
1	1	2	3	4	5	6	7	8	9	10	11				
2	2	3	4	5	6	7	8	9	10	11	12				
3	3	4	5	6	7	8	9	10	11	12	13				
4	4	5	6	7	8	9	10	11	12	13	14				
5	5	6	7	8	9	10	11	12	13	14	15				
6	6	7	8	9	10	11	12	13	14	15	16				
7	7	8	9	10	11	12	13	14	15	16	17				
8	8	9	10	11	12	13	14	15	16	17	18				
9	9	10	11	12	13	14	15	16	17	18	19				
10	10	11	12	13	14	15	16	17	18	19	20				

Finding Patterns in Numbers

Keith Devlin is a Stanford mathematician who wrote a beautiful book titled *Mathematics, the Science of Patterns: The Search for Order in Life, Mind and the Universe* (1994). I highly recommend it. In it Keith argues that mathematics is not a subject of rules and procedures but is all about the study of patterns. I agree with this and find that when students change their idea of their role in mathematics—from being a rule follower to a pattern seeker—everything changes for them.

The world is filled with beautiful patterns that can be found every day, and I choose to find and appreciate patterns whenever I can. Hotels are my latest place for appreciating patterns—have you ever noticed how the carpets and floors in hotels are almost always made up of patterns? And that the wall art is often pattern based too? But I also see patterns in all of mathematics. For example, if I ever need to multiply an even number by 5, I can divide the number by 2 and multiply it by 10; 18×5, for example, is the same as 9×10. This is a pattern that always works, in the same way that the formal algorithm for multiplication that involves "carrying" numbers is a different (less beautiful!) kind of pattern that always works. I conducted an interview with another important mathematician—Francis Su—who talked in the interview about the joy of pattern seeking and that when we teach students to explore patterns and to value number flexibility, we give them an approach that has lifelong value. You can watch the whole interview with me and Francis on the Youcubed page called "Thought Leaders" (https://www.youcubed.org/resource/21st-century-mathematics/). Francis has also written a lovely book called *Mathematics for Human Flourishing*, in which he describes the benefits of pattern seeking.

In an important study of first-grade children, Kidd et al. (2014) found that those who were taught patterning, compared with others who were taught reading or mathematics, improved their achievement in reading and mathematics by six to eight months. To me, this study hints at the power of a pattern-seeking perspective for students. In this set of activities, we aim to encourage that perspective for your students, with the particular pattern-seeking activity Francis recommends in his interview with me—working with the hundred chart. We have created a set of activities around the hundred chart so that students start to see it as a pattern playground and to notice and wonder about patterns.

In our Visualize activity, we ask students to explore what they see in the chart. This activity then moves to exploration with puzzle-piece types of patterns that have some blanks for students to fill in. Students will need to think about the larger hundred chart and determine the numbers that fill in the puzzle spaces to complete the pattern. The puzzles are created so that students can think about 1s, 5s, and 10s, moving up and down or left and right. I love these puzzles because they are fun, they encourage students to figure out patterns, and they ask students to share their strategies about how they saw the patterns and figured them out.

In the Play activity, students learn to skip-count and to notice patterns in the chart by skipping by 2s, 3s, 5s, and other numbers. We also share how students can use snap cubes to create interesting visuals. In the extension, students can read numbers to other students and ask them to "guess their pattern."

In the Investigate activity, we connect mathematics to art. Skip counting in a hundred chart can look like a woven basket. We connect patterns to weaving using a particular style called Pomo basket weaving. Students are encouraged to use my favorite math manipulative—Cuisenaire rods—as a way to add color, and we ask students to color-code their counting patterns on the hundred chart as if they were using the colored rods. The patterns wrap around, which creates an effect similar to the weaving patterns of Pomo baskets—which is very cool!

Jo Boaler

References

Devlin, K. (1994). *Mathematics, the science of patterns: The search for order in life, mind and the universe*. New York, NY: Scientific American Library.

Kidd, J. K., Pasnak, R., Gadzichowski, K. M., Gallington, D. A., McKnight, P., Boyer, C. E., & Carlson, A. (2014). Instructing first-grade children on patterning improves reading and mathematics. *Early Education and Development, 25*(1), 134–151. doi:10.1080/104 09289.2013.794448

Su, F. (2020). *Mathematics for human flourishing*. New Haven, CT: Yale University Press.

Patterns in the Hundred Chart

Snapshot

Students explore patterns in the hundred chart, opening inquiry into 10s and 1s, place value, and counting.

> **Connection to CCSS**
> 1.NBT.2, 1.NBT.1, 1.NBT.5, 1.OA.5

Agenda

Activity	Time	Description/Prompt	Materials
Launch	10 min	Show students the Hundred Chart sheet and ask, What do you notice? What is this chart showing? Give students a chance to turn and talk and make observations. Note or color-code some of what students see.	• Hundred Chart sheet, to display • Colors
Explore	25–30 min	Partners look for patterns in the hundred chart, noting and color-coding those they find.	• Hundred Chart sheet, at least one per partnership • Colors
Discuss	10–15 min	Discuss the patterns that students found and how to describe them. Discuss what patterns emerge from moving in different directions on the chart and why these patterns might exist. Press students to use increasingly precise language for 1s, 10s, and their movements on the chart.	• Hundred Chart sheet, to display • Colors
Extend	20–30 min	Partners figure out the missing numbers on the Hundred Chart Puzzle sheets using the hundred chart as a reference or by reasoning about the patterns it contains. Discuss the strategies students developed and how they used patterns to solve these puzzles.	• Hundred Chart Puzzle sheets, per partnership • Make available: Hundred Chart sheet

To the Teacher

In this activity, we invite students to explore the patterns found in the hundred chart. Students may have encountered a hundred chart in kindergarten or in your classroom space, where it is often used to support the connection between numerals and oral counting sequences. As students extend their counting capacity toward and beyond 100, they are poised to make sense of the patterns in our number system, particularly place value. The hundred chart organizes numbers so that students can see the repetitions in the 1s and 10s places and reason about what those patterns represent. For instance, as you move horizontally on the hundred chart, the numbers increase (or decrease) by 1, and we see a change in the 1s place. Moving vertically, the numbers increase (or decrease) by 10, and we see a change in the 10s place only, while the 1s place remains the same. Other patterns emerge if you move in different ways, such as counting by 2s or moving diagonally. Encourage students to make connections between the ways they move across the chart, the patterns that emerge, and the concept of 10s and 1s. Color coding and labeling offer a visual way of accessing, identifying, and describing these patterns.

In the extension, we invite students to use these patterns within the hundred chart to reason about the relationship between numbers and their positions. Students may want to refer directly back to the hundred chart to solve these puzzles, and we encourage you to let them. As they work, you can promote thinking about patterning and number relationships by asking questions about how students could solve the puzzles if they did not have the hundred chart.

Activity

Launch

Launch the activity by showing students the Hundred Chart sheet on the document camera and inviting students to make observations. Ask, What do you notice? What is this chart showing? Give students a chance to turn and talk to a partner about what they see. Invite students to share their observations, coming up to the hundred chart to point out what they notice. Students may notice some patterns or make observations about how the chart shows counting from 1 to 100. You might make notes or use colors on the Hundred Chart sheet to show these observations. Tell students that this chart is full of patterns and that their goal today is to be a pattern seeker.

Explore

Provide partners with a Hundred Chart sheet and colors. Students explore the following questions:

- What patterns do you notice?
- What patterns can you find by moving from square to square in the chart?
- How can you describe the patterns you see?

Students use color and words to label the chart, showing the different patterns they see. Partners may want multiple copies of the Hundred Chart sheet to help them keep track of different patterns.

Discuss

As a class, discuss the following questions:

- What patterns did you notice?
- What happens as you move in the chart? (Focus on patterns with 10s and 1s.)
- How can you describe the patterns you see?
- Why do you think these patterns exist?

You might have a partnership come up and point out a set of numbers that makes a pattern, and then have the class try to figure out what the pattern is. You could ask, What could this pattern be?

Throughout the discussion, press students to both show and describe the patterns they have found, and encourage students to use increasingly precise language for 1s, 10s, skip counting, counting by __s, and the direction of movement.

Extend

Provide partners with the Hundred Chart Puzzle sheets and access to the complete hundred chart. Students try to figure out what numbers are missing from each section of the hundred chart shown in the puzzle. Students can use the hundred chart as a support, or they can develop strategies for figuring out the missing values by using the relationships and patterns they noticed in the table. As students are solving and afterward in discussion, ask, How can (or did) you figure out what number was missing? How could (or did) you figure it out without looking back at the hundred chart? What patterns did you use? What was your reasoning?

Look-Fors

- **Are students noticing patterns horizontally, vertically, and diagonally?** Different patterns emerge when you move in different ways across the hundred chart. We want students to have opportunities to notice and compare horizontal and vertical patterns, and see how these patterns merge when moving diagonally. If you notice students attending only to vertical patterns, for example, ask questions to prompt looking in a new way. For instance, you might say, "I see all the patterns you found moving up and down in the hundred chart. What other ways could you move on this chart? What happens if you move a different way?"

- **Are students describing the patterns using any place value language (10s and 1s) or only by noticing digits (first number or second number)?** As students begin to notice patterns in the hundred chart, they will need language to articulate what they see. It is natural that students begin by referring to the digits in the numerals they attend to as the "first number" and "second number" well before they may use the language of 10s and 1s. When you hear this language, ask some questions about what it means. You might clarify by asking, When you say the "first number," are you talking about this digit here? while pointing to the 10s place. This can support students in taking up the term *digit*, and you can then ask, What does this digit mean? Through this kind of conversation, you can connect numbers to digits to place value, so that students can ultimately begin to see and speak about 10s and 1s.

- **If students are tackling the puzzles, are they using patterns to find the missing value?** Although it is entirely acceptable for students to use the hundred chart as a reference for navigating these puzzles, the goal is for them to use the patterns they have noticed in the chart to reason about numbers in relation to one another. Ask students questions to uncover whether they used the chart as a map or reasoned about patterns when solving. Ask, How did you know what number went in this blank? When students confront blank boxes connected diagonally only, are they able to think through the vertical and horizontal movements that combine into a diagonal motion? Similarly, movement forward is easier to reason about than movement backward. How do students reason about moving up or to the left on the chart? Are students able to count backward by 1s and 10s? These are big ideas that students are actively developing, and they will benefit even more from thinking about how to move backward through the number system than they will moving forward. Support students in counting back through oral counting and referring back to the hundred chart when they need support.

Reflect

What happens when you move down a column in the hundred chart? Why?

 **Hundred Chart**

1	2	3	4	5	6	7	8	9	10
11	12	13	14	15	16	17	18	19	20
21	22	23	24	25	26	27	28	29	30
31	32	33	34	35	36	37	38	39	40
41	42	43	44	45	46	47	48	49	50
51	52	53	54	55	56	57	58	59	60
61	62	63	64	65	66	67	68	69	70
71	72	73	74	75	76	77	78	79	80
81	82	83	84	85	86	87	88	89	90
91	92	93	94	95	96	97	98	99	100

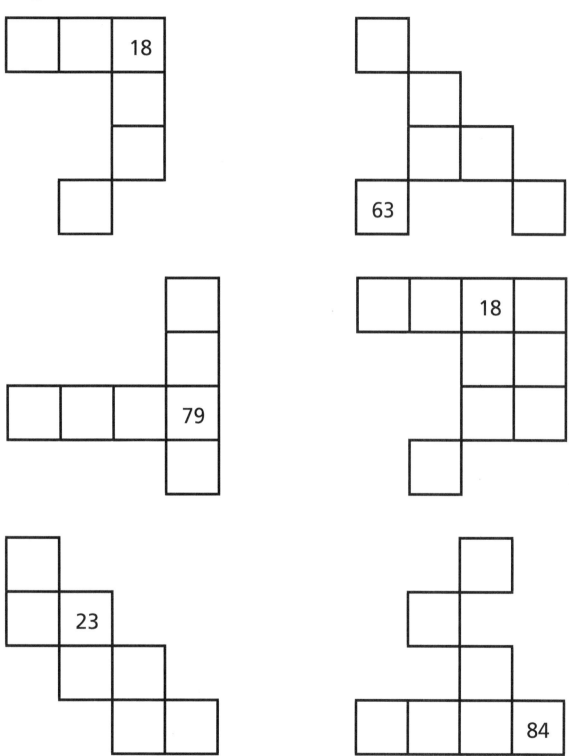

Mindset Mathematics, Grade 1, copyright © 2021 by Jo Boaler, Jen Munson, Cathy Williams.
Reproduced by permission of John Wiley & Sons, Inc.

Skipping across the Hundred Chart

Snapshot

Students play with the patterns created by skip-counting across the hundred chart, exploring both visual and numerical patterns.

Connection to CCSS
1.NBT.1, 1.NBT.2, 1.OA.5

Agenda

Activity	Time	Description/Prompt	Materials
Launch	10 min	Show students how to skip-count by 2s on the hundred chart and invite them to notice the patterns that skipping creates in the chart.	• Hundred Chart sheet, to display • Highlighter
Play	30+ min	Students choose a skip-counting pattern to color on the hundred chart and then explore the patterns they can see with their eyes and those they notice in the numbers colored. Partners can play with new patterns.	• Hundred Chart sheet, multiple per partnership • Colors
Discuss	10–15 min	Invite different groups to share the patterns they created and ask the class to describe how the groups counted, the patterns they can see, and the patterns they notice in the numbers. As groups share, compare the patterns created and name this process as *skip-counting*.	

Activity	Time	Description/Prompt	Materials
Extend	20–35 min	Partners play a game of Guess My Pattern. Partners each have a hundred chart and tools for temporarily marking patterns. One partner creates a pattern and then reads numbers aloud to their partner until the partner can guess the next number and the pattern being made.	• Hundred Chart sheet, per student • Tools for marking, such as bingo chips or counters • Barrier, such as a file folder • Optional: laminated Hundred Chart sheets and dry-erase markers

To the Teacher

In this activity, we extend the work students started in the Visualize activity exploring the hundred chart by playing with skip-counting patterns. The hundred chart enables students to see how skip counting works by showing both the numbers that are counted and those that are skipped. Because students can see both of these groups, they can also see with their eyes patterns in how these numbers are arranged in space. For instance, counting by 2s, 5s, and 10s creates columns of numbers counted and those that are not. Counting by 3s or 4s, by contrast, creates diagonal, zigzagging patterns. Students can connect these visual patterns to numerical patterns. For instance, the column pattern of skip-counting by 2s also means that all of

Hundred chart patterns showing skip counting by 5s in yellow and by 6s in blue the numbers ending with the digits 2, 4, 6, 8, and 0 get counted, but the zigzagging pattern of counting by 4s creates a very different numerical pattern, with a cycle of digits in the 1s place (4, 8, 2, 6, and 0, repeating). Connecting the visual to the numerical is at the heart of the patterning work in this activity.

To make this connection between the physical space of the hundred chart and numerical patterns, we encourage you to find ways for students to interact with the chart with their bodies. If you have an outdoor space, we recommend drawing a large hundred chart on the pavement—either temporarily with chalk or permanently with paint. Students can use the interactive hundred chart for this activity, marking numbers with cones or bean bags and literally skipping across the chart. They can also use it to create their own playground games, in the spirit of hopscotch and four corners.

Activity

Launch

Launch the activity by showing students how to skip across the hundred chart on the document camera. As you skip-count by 2s, say each number on the chart aloud, but only highlight or color the multiples of 2. When you get to the end of the row, highlighting 10, ask the class, What do I do now to keep going with this pattern? Give students a chance to turn and talk to a partner, and then invite students to come up and show how to wrap around the row to continue.

After you have continued with the skip counting for at least three rows, ask, What patterns do you see when we skip across the hundred chart? Give students another chance to turn and talk to a partner about what they notice. Invite students to share the different patterns they see, coming up to the chart to point these out. Make sure that students notice that you have skipped in the same way consistently across the chart, and that this skipping has created patterns. For instance, the colored squares all line up in columns, and those columns each have numbers ending in the same digit.

Play

Provide partners with access to multiple copies of the Hundred Chart sheet and colors. Partners decide how they want to skip (by 2s, 3s, 4s, 5s, or something else) to explore the patterns that skipping creates. Students then skip-count and color in the steps they take across the chart. Partners then use the chart they have created to explore the following questions:

- What patterns do you see with your eyes?
- What patterns do you notice with the numbers?
- What do your patterns make you wonder?

Encourage students to label the patterns they see, both the visual patterns (such as columns or diagonal lines) and the numerical patterns (such as numbers ending in the same digit or sequence of digits) created.

Partners can explore what happens with a different skipping pattern, answer the same questions, and then compare the patterns they made using the following questions:

- How are the patterns you made similar?
- How are they different?

Discuss

Invite a partnership to show their pattern on the document camera, and ask the class to discuss the following questions:

- What is the pattern? How did they skip?
- What patterns does it make?
- What would come next in the pattern if our chart were even larger?
- Does anyone have a different kind of pattern? Or one that is similar but not the same?

For each group that shares, discuss the same questions about the new patterns and how they compare to previously shared patterns. During the discussion, name this process as *skip counting*, where some numbers are skipped and others are counted, following a pattern. Point out examples of "skip-counting by 2s" and other numbers so that students hear how this process can be described mathematically.

Extend

Play a game of Guess My Pattern. Provide partners with one Hundred Chart sheet each and a way to mark numbers, such as tinted bingo chips or counters. Alternatively, the hundred charts could be laminated or put in sheet protectors and students use dry-erase markers. Partners sit across from one another with a barrier, such as a file folder, to block the view of each other's charts.

Guess My Pattern Rules

- Partner A creates a pattern on their own hundred chart.
- Partner A reads aloud the first few numbers in their pattern to their partner.
- Partner B marks the numbers on their own hundred chart to keep track of the pattern.
- Partner B can ask for more numbers to be read aloud, until they can make a guess of what number comes next and what the pattern is.
- When Partner B can predict the next number and describe the skip-counting pattern (such as "skip-counting by 4s"), players swap roles.

After students have had a chance to play a few rounds, discuss the following questions:

- What information did you need before you could tell what the pattern was?
- What clues were useful?
- How did you make your predictions about what number came next?

Look-Fors

- **Are students paying attention to place value?** As students look for patterns in the numbers marked on their hundred chart, our aim is that they connect the numbers they see to place value. Many of the patterns that emerge encourage students to attend to the 1s place or 10s place only to describe the pattern, as when counting by 5s and all of the numbers in the pattern have a 5 or 0 in the 1s place. However, we don't want students to reduce this pattern to "numbers that end in 5 and 0," which treats the numbers as free-floating digits. Instead, encourage students to use the language of place value by revoicing that the pattern shows, for example, "all numbers with 5 or 0 in the 1s place" or "all numbers with five or zero 1s." Ask students questions about why they think this pattern works in this way, such as, Do you think this pattern will continue? Why or why not? Why does it always come back to numbers with these digits in the 1s place? What's happening? Students may find that modeling the pattern with blocks helps make sense of what is going on.

- **How do students understand patterns that appear diagonally on the chart?** Counting by 2s, 5s, and 10s may lead students to think that skip counting always generates a simple, rhythmic pattern. But when counting by 3s, 4s,

6s, or many other numbers, the patterns become more complex, the sequences longer, and the visual pattern diagonal or zigzagging. Of course, there is still a pattern to these numbers, but students may need to see even more of it to make sense of what is happening or to perceive the pattern. If students are tempted to say there is no pattern, that the numbers are random, encourage them to look at it again. They may want to post it on the wall and stand back from it to see it more clearly, and students who highlight the numbers without much color may want to go back and color the squares in more fully to better see what is happening. Other students may want to say the numbers aloud to hear the pattern. Still others might benefit from looking down different columns and looking for patterns in each column before taking in the whole. Give students lots of ways to search for patterns, and if they are still struggling, bring the chart to the class and ask the group to grapple with it together.

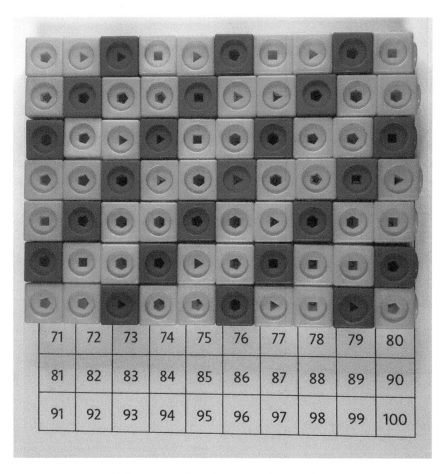

71	72	73	74	75	76	77	78	79	80
81	82	83	84	85	86	87	88	89	90
91	92	93	94	95	96	97	98	99	100

Skip-counting by 3s using cubes

- **Do students need to build the numbers with blocks to see how they wrap around the chart?** Another way of visualizing these patterns is to construct a sequence with colored snap cubes that can be broken to represent the lines of the chart. For example, if students are trying to make sense of the patterns in skip-counting by 3s, they could make a long snap cube string using two blue cubes then one yellow cube, repeatedly for 30 or 40 cubes. This string would show clearly the consistency in the pattern and how its rhythm plays out along a long number sequence. Then students can break the string every 10 cubes, laying the groups of 10 below one another, recreating the hundred chart. This way of modeling the pattern draws attention to where the pattern breaks or wraps around each group of 10, making a consistent pattern suddenly look more chaotic. Encourage students grappling with these patterns to try out this form of modeling to make sense of why diagonal patterns emerge when skip-counting by some numbers but not others.

Reflect

What were the most interesting patterns you found while skip-counting on the hundred chart? What other ways of skip-counting would you like to explore? Why?

Weaving Patterns

Snapshot

Students investigate the connection between patterns, weaving, and numbers by creating a sequence with Cuisenaire rods that they weave onto the hundred chart.

Connection to CCSS
1.OA.5, 1.NBT.1, 1.NBT.2

Agenda

Activity	Time	Description/Prompt	Materials
Launch	10–15 min	Show students the Pomo Basket sheet and tell them that creating such a design involves many mathematical ideas. Ask, What patterns do you see? Invite students to talk to one another and share their observations. Tell them they will be investigating how to make woven patterns on the hundred chart using Cuisenaire rods. Introduce these if students are not yet familiar with them. Show students a pattern with Cuisenaire rods and discuss how to transfer it to the Centimeter Hundred Chart.	• Pomo Basket sheet, to display • Cuisenaire rods, to display • Colors • Centimeter Hundred Chart sheet, to display
Explore	30+ min	Partners create repeating patterns with Cuisenaire rods, such as AB or AAB, and then figure out how to transfer these to the Centimeter Hundred Chart. Using this representation, students investigate the patterns created by weaving the rods across the hundred chart.	• Cuisenaire rods, per partnership • Colors, per partnership • Centimeter Hundred Chart sheet, multiple per partnership

Activity	Time	Description/Prompt	Materials
Discuss	15–20 min	Post students' weaving patterns and do a gallery walk with partners. As students examine each pattern, they consider what the rod pattern is, the patterns it creates on the hundred chart, and which they find most interesting or challenging. Discuss students' observations.	• Students' weaving patterns • Tape or magnets for posting
Extend	30+ min	Explore other weaving patterns in multiple representations, such as baskets, woven rugs, fabric, tiling, or wallpaper. Invite students to figure out what the repeating pattern is and how it could be translated into numbers or Cuisenaire rods.	• Resources with weaving patterns, such as textiles, books, or images or videos from the internet • Make available: Cuisenaire rods and Centimeter Hundred Chart sheets

To the Teacher

In this investigation, we take a big leap in exploring the hundred chart in a new way. In the Play activity, we examined the patterns created by skip-counting across the hundred chart. Students likely discovered many different ideas by skip-counting in different ways and thinking about both visual patterns and numerical patterns. Here, we extend these ideas by thinking about weaving across the chart, rather than skipping. We were inspired by weaving patterns in indigenous basketry, where patterns spiral around, overlapping in interesting ways. In the basket created by a Pomo artisan that we show in this activity, dark and natural fibers are alternated with a long string of natural knots, then individual dark and natural knots, and then a final long string of dark knots. This pattern of light and dark, when wrapped around the basket, creates a visual pattern of elongated parallelograms.

We were fascinated by how a sequence can create interesting visual patterns when wrapped, and we realized that a similar phenomenon would occur if we wrapped a sequence across the hundred chart. To do this, we have crafted an investigation using Cuisenaire rods and a hundred chart that is scaled to fit them. Interesting visual patterns emerge when you wrap a Cuisenaire rod pattern sequence across

Pomo basket with a parallelogram pattern
Source: 1025870 Paul Marcus/Shutterstock.com

the hundred chart. For example, a sequence of light green, which is 3 units long, and purple, which is 4 units long, alternating in an AB pattern, creates the stair-step pattern when wrapped around the hundred chart as shown. The staggered stairs in this pattern appear to climb from the lower left to the upper right. By contrast, the ABC pattern of red (2 units long), dark green (6 units long) and yellow (5 units long) makes a stair-step pattern with floating red stairs that appear to climb in the opposite direction, from lower right to upper left. Isn't that interesting?

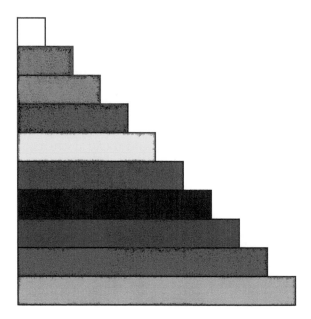

A set of typical Cuisenaire rods

Mindset Mathematics, Grade 1

1	2	3	4	5	6	7	8	9	10
11	12	13	14	15	16	17	18	19	20
21	22	23	24	25	26	27	28	29	30
31	32	33	34	35	36	37	38	39	40
41	42	43	44	45	46	47	48	49	50
51	52	53	54	55	56	57	58	59	60
61	62	63	64	65	66	67	68	69	70
71	72	73	74	75	76	77	78	79	80
81	82	83	84	85	86	87	88	89	90
91	92	93	94	95	96	97	98	99	100

1	2	3	4	5	6	7	8	9	10
11	12	13	14	15	16	17	18	19	20
21	22	23	24	25	26	27	28	29	30
31	32	33	34	35	36	37	38	39	40
41	42	43	44	45	46	47	48	49	50
51	52	53	54	55	56	57	58	59	60
61	62	63	64	65	66	67	68	69	70
71	72	73	74	75	76	77	78	79	80
81	82	83	84	85	86	87	88	89	90
91	92	93	94	95	96	97	98	99	100

Two different Cuisenaire rod patterns on a hundred chart

Each of these visual patterns has a corresponding numerical pattern, though many are far more complex than first graders typically grapple with. Our aim is to connect the hundred chart to patterns in the real world, many of which are visual. There is no need for students to be able to analyze the roots of each of the patterns they create. Rather, we hope to inspire curiosity and connections.

The extension takes this further, by inviting you to investigate similar wrapping patterns in other visual art forms, such as wallpaper, rugs, fabric, or tiling. We have left this very open as a true investigation, in which you might provide students with resources, such as images from the internet, books of wallpaper, or fabric samples, for them to explore. Students might figure out what the patterns are, how they wrap, and how the patterns could be represented with the tools they have, Cuisenaire rods and the hundred chart. They might also design their own wrapping patterns to create a new fabric, rug, or wallpaper.

Activity
Launch

Launch the activity by showing students the Pomo Basket sheet on the document camera. Tell students this is a basket woven by the Pomo people who are indigenous to Northern California and that creating this and other basket patterns involves lots of mathematical ideas. Ask, What patterns do you see? Give students a chance to turn and talk to a partner, and then invite students to come up and point out the patterns they notice. Press students to use precise language to describe the pattern, such as *vertical*, *horizontal*, *diagonal*, and *repeating*. Be sure students see that the basket is woven in a wrapped spiral using different colored material.

Tell students that they are going to explore how to make weaving patterns on the hundred chart using Cuisenaire rods to be the bands of color like those in the Pomo basket. If your students have not seen Cuisenaire rods before, provide students access to the pieces and ask them to make observations about the rods. Students should notice that they come in different lengths and that they can be arranged in increasing (or decreasing) lengths that can be counted.

On the document camera, make a simple pattern using the Cuisenaire rods, such as purple-red, repeating for several iterations. Tell students that you can use this sequence to weave a pattern on the hundred chart. Show students the Centimeter Hundred Chart sheet and ask, How could we put this pattern on the chart? Give students a chance to turn and talk to a partner, and then invite students to share ideas, coming up to the document camera to show how they might place these blocks onto the hundred chart. Show students how they could use colors to mark your pattern, such as purple (4 units long) and then red (2 units long), on the hundred chart. Only color within the first row and do not show students how to wrap around from one row to the next. If students ask, then pose the question to the class, What could we do at the end of the row?

Explore

Provide partners with Centimeter Hundred Chart sheets, colors, and Cuisenaire rods. Partners first make a pattern with rods and lay them out end to end in a line. Patterns can be simple, such as A or AB, or more complex, such as AAB or ABC. Students do not need to know the names for the patterns they create; rather they should use the rods to design a pattern that repeats. They may describe it simply as "orange-brown" or "red-red-yellow." Then partners explore the following questions:

- How can you transfer this pattern to the hundred chart? Use colors to help you record your patterns on the chart.
- What can you do when you come to the end of a row? How can you still continue the pattern?
- Once you have colored your patterns on the hundred chart, what do you see?
- Do you notice any patterns with the numbers?
- What are you wondering about weaving patterns? What else can you try?

Encourage students to try different Cuisenaire rod patterns to see what kind of weaving patterns they create. Student explore the question, What kinds of rod patterns make the most interesting weaving patterns?

Discuss

Post students' weaving patterns on the walls or on table tops. Do a gallery walk with partners, moving slowly from pattern to pattern. They may not look at all of the patterns created; it is better that they closely examine a few. As they walk, ask students to look for answers to the following questions and then discuss them:

- How is each pattern constructed? What is the pattern?
- What kinds of patterns with rods and numbers make diagonal patterns?
- What kinds of patterns with rods and numbers make vertical patterns?
- Which patterns did you find most interesting? Why?
- Were there any patterns that didn't look like patterns? Why or why not?
- Which patterns were challenging to figure out? Why?

Extend

Where else do these kinds of weaving patterns show up in our world? Encourage students to explore patterns that repeat and wrap around in artifacts such as baskets, woven rugs, fabric, tiling, or wallpaper. You might watch videos of people weaving with fibers to see how these are constructed. Explore the question, How could these patterns be represented with rods or numbers?

Look-Fors

- **How are students thinking about wrapping their sequence around the end of a row on the hundred chart?** With our Centimeter Hundred Chart sheet, students can lay their Cuisenaire rods directly on the chart to help them transfer the sequence. But most sequences that students create will not end precisely at 10; that final rod is likely to hang off the end of the chart, raising the question of what to do about the extra length that does not fit. Pay attention as students get to this place, and ask questions to help them think about how to approach wrapping the pattern around. Thinking it through will be easier if students can quantify, using units, how much of the rod fits on the row and how much hangs off. For instance, if students have a yellow rod where they can say 3 of the units fit onto the row and 2 units hang off, you can ask, Where are those 2 units? What numbers would they be? Where are those numbers? Ask, How can you wrap this extra bit around, as if it were a basket? Stay with students as they grapple with this idea and then have them think through their solution by asking, Why does that make sense? How do you know you represented the full rod, even though it is partly on this row and partly on the next?

- **Are students able to think about both the pattern sequence and the patterns created by wrapping?** One of the fascinating things about this activity is that it involves four different kinds of patterns. First, there is the sequence of rods that students create, such as red-white-red-white-red-white, and so on. This pattern can be shown as a never-ending repeating sequence of rods arranged linearly. Second, these colors also represent numbers, so in the case of red-white there is also a numerical pattern of 2-1-2-1-2-1, and so on. When this gets translated onto the hundred chart, new visual patterns emerge with staggering, shapes, stairs, and spots. Finally, these shapes contain numerical patterns also. The red-white pattern, for instance, could be described by the final number in each colored group: 2, 3, 5, 6, 8, 9, 11, 12, and so on. Here a numerical pattern exists as pairs of consecutive numbers and then skipping a number. There are just lots and lots of different sorts of patterns that come from the initial sequence of rods. Help students think about all of these and make connections between them, wherever possible. Students may struggle with language to specify where the different patterns come from, so help them name when they are noticing patterns with their eyes (colors and shapes) or with the numbers, and whether they see patterns in the rods or on the hundred chart. Ask questions about why and how students think that the patterns are related. Again, it is not important that students find answers to all these questions, but rather that they try to explore connections.

- **Do students notice any numerical patterns?** The numerical patterns are the trickiest to see, because they are far less visual than those involving colors and shapes. Students are very likely to focus first on what their eyes see, and we encourage students to look for patterns wherever their eyes are drawn. However, they may simply overlook numerical patterns, unless you or others invite them to look. First, do you notice any students talking about the numbers, either as represented by the lengths of the rods or as shown on the hundred chart? If so, ask those students to talk more about what they see, what they think it means, and what they wonder. You may even invite them to point out these observations to the class partway through the investigation. If you do not hear students attending to the numbers, ask questions that remind them that this is an avenue to explore. You might say, "I wonder if there are any patterns in the numbers in your stairs" or "One way to describe your pattern is blue-yellow, but these are not the same length, are they? Is there a way to describe this pattern using numbers instead of colors? What happens to those numbers on the hundred chart?" The goal is simply to encourage students to look for any patterns they can find.

- **Are students wondering?** One of the central aims of this investigation is to inspire students to see patterns in new ways, connect seemingly different kinds of patterns to one another, and wonder about how patterns come to be. Listen to students as they make sequences of rods, lay these on top of their hundred chart, and observe the patterns that emerge from wrapping. Do you hear notes of surprise in their voices? Do you hear questions? We hope that students find these interesting, intriguing, and inspiring. Cultivate an atmosphere of wondering by noticing and revoicing students' wonders and by posing your own. You might see a pattern over a child's shoulder and just ask, I wonder why it slants that way? and walk away. Not every question can or needs to be answered. The power is in asking the questions and knowing you can.

Reflect

Where have you seen woven patterns before? What mathematical ideas do you think the weaver was thinking about to make those patterns?

Centimeter Hundred Chart

1	2	3	4	5	6	7	8	9	10
11	12	13	14	15	16	17	18	19	20
21	22	23	24	25	26	27	28	29	30
31	32	33	34	35	36	37	38	39	40
41	42	43	44	45	46	47	48	49	50
51	52	53	54	55	56	57	58	59	60
61	62	63	64	65	66	67	68	69	70
71	72	73	74	75	76	77	78	79	80
81	82	83	84	85	86	87	88	89	90
91	92	93	94	95	96	97	98	99	100

Using Place Value to Add and Subtract

A key idea in early mathematics that underpins place value understanding is grouping numbers and dealing with them as units (Hiebert & Wearne, 1992). The learning of partitioning of a continuous quantity or a set of discrete objects into equal-size subparts is extremely important, and it underpins later mathematics learning, including fraction learning. For many students, difficulties in algebra can be traced back to an incomplete understanding of number partitioning (Hiebert & Wearne, 1992). Ideally, students will experience number groupings with many different representations, such as snap cubes, Cuisenaire rods, and more. Different representations highlight different aspects of grouping structures, which help students build essential brain connections. In a study conducted by Hiebert and Wearne with first-grade students, the researchers contrasted learning place value in a typical way with learning place value conceptually, which showed that the conceptual teaching advanced students' understanding significantly.

The activities in this big idea, which engage students conceptually, extend the activities in Big Idea 3, which focused on representing and modeling situations. Now we connect these modeling ideas to patterns using the hundred chart. Rather than having students think of numbers as individual words or digits, we help them learn to see larger numbers as collections of 10s and 1s, which can be taken apart and put together. An important goal of these activities is for students to consider mathematical situations as a whole, to develop meaning and understanding. This is the opposite of the often-used approach of focusing on key words—where students are trained to perform an operation when they see a particular word, which leads to considerable mathematical difficulties. An example of this approach at work would be students seeing a problem such as 100 minus 4—interpreting "minus" as needing to line up a subtraction and then perform a difficult algorithm. Instead, students should be encouraged to consider the whole situation and see that there are many more valuable methods than performing an algorithm. This big idea focuses on reading and understanding situations, and students talking to each other about them.

In our Visualize activity, students use concrete models to learn about decomposing. Using a set of problems all about recess, students explore how 10s and 1s can be used to join numbers and how concrete models can show decomposing using place value. We encourage students to think, and then invite them to share their answers and strategies. Teachers can highlight strategies that use 10s and 1s, particularly those that *decompose* 10s and 1s to join. We also ask students how they could use cubes to model this way of thinking and show their ideas with cubes.

In our Play activity, students use their own ideas to model and solve problems involving numbers between 20 and 100. We think students could have playful fun with cubes, rekenreks, number lines, or other models. The learning is about interpreting the meaning of problems, which is often overlooked in textbooks. Textbooks often move students quickly to number problems without words or context, and the problems do not involve action, such as giving and taking. Here the problems are more complex in wording and structure, with examples such as picking flowers and starting to jump rope. This gives students the opportunity for challenge, something we want for all students, as brains grow most when students are struggling—a good message to keep reminding the students about and to remember for ourselves.

In our Investigate activity, students explore composing and decomposing with 100 coins, connecting addition, skip counting, and place value. It is always a good idea to help students recognize different coins and not assume they will know their values. When my family moved to the US from England, my daughter was lost in her first math class, as it was using money, and she did not know the value of a dime, a quarter, or a dollar. (She did know pounds and pennies, though!) This made me realize that this must be a common experience for many students, and not one we want them to have. A simple chart on the wall showing the different values can help with this. The task begins with a question that we think will grab your students: If I have 100 cents in my pocket, what coins might I have? We love this question, as there are many possible solutions—the best kind of maths problem!

Jo Boaler

Reference

Hiebert, J., & Wearne, D. (1992). Links between teaching and learning place value with understanding in first grade. *Journal for Research in Mathematics Education*, 98–122.

Recess!

Snapshot

Using a set of problems all about recess, students explore how 10s and 1s can be used to join numbers and how concrete models can show decomposing using place value.

> **Connection to CCSS**
> 1.NBT.4, 1.NBT.2, 1.OA.1

Agenda

Activity	Time	Description/Prompt	Materials
Launch	10–15 min	Do a number talk: 15 + 5. Give students think time and then invite them to share their answers and strategies. Highlight strategies that use 10s and 1s, particularly those that *decompose* 10s and 1s to join. Ask students how they could use cubes to model this way of thinking, and show students' ideas with cubes.	• Space to record a number talk, such as a whiteboard or chart paper • 20 snap cubes, arranged in a stick of 15 and a stick of 5
Explore	30 min	Provide students with the Play Ball! sheet and access to the other tasks in the Recess Problem Set. Students use snap cubes, drawings, and numbers to invent ways of solving these problems. Students consider how 10s and 1s can help them solve; as you confer, highlight strategies that use place value. Students represent their thinking on paper.	• Play Ball! sheet, per partnership • Snap cubes, at least 50 per partnership, or access to a large shared collection • Make available: Kids on the Playground, A Pile of Jackets, and Playing Soccer sheets, per partnership

Activity	Time	Description/Prompt	Materials
Discuss	10–15 min	Discuss students' invented strategies, both those that involve direct modeling and those that use place value. Chart and name the different strategies that students invented and make connections between similar strategies. Discuss how 10s and 1s were helpful to those who used decomposing strategies.	• Chart and markers • Snap cubes
Extend	30+ min	Brainstorm recess situations in your school that would involve joining (or separating) and invite students to create and solve their own problems. Discuss the strategies students used, how they were connected to those shared earlier, and any new ideas that emerged.	• Make available: snap cubes • Chart and markers

To the Teacher

This activity, and the entire big idea, is an extension of earlier work in Big Idea 3 on representing and modeling joining and separating situations. Here, we connect these modeling ideas to patterns using the hundred chart that students investigated in Big Idea 7. Rather than having students think of numbers as individual words or digits, we want them to see larger numbers as collections of 10s and 1s, which can be taken apart and put together in any way that is useful to the mathematician. As with earlier activities, we build on the work of *Children's Mathematics: Cognitively Guided Instruction* (Carpenter, Fennema, Franke, Levi, & Empson, 2015), which we highly recommend.

We launch the activity with a number talk, 15 + 5, that encourages students to make a 10 and decompose by place value. If you think this particular number talk will not challenge your students, you can select different numbers that encourage the same kind of thinking, such as 21 + 9 or 34 + 6. In discussing this number talk, focus on strategies that use 10s and 1s and be sure that students see that this is a useful way of thinking. This is not to say that other strategies shouldn't be valued; any strategy based on reasoning is valuable and offers students ways of making connections and comparisons. You may see students using direct modeling strategies, those that involve showing and counting each individual item in the task. Encourage students to use the strategies most comfortable to them, and then try something new with the same task.

We have provided a four-task Recess Problem Set, in which all the stories are join—result unknown. We selected situations that we thought would be familiar to many students, but consider writing your own tasks so that you can modify names, contexts, and the numbers involved to better match your students' experiences. We've focused on numbers that encourage students to decompose and make new 10s, particularly those where the 1s are trains of 10 (5 and 5, 6 and 4, 7 and 3, 8 and 2, or 9 and 1), and if you write your own tasks, we encourage you to use this as a guideline for number selection.

Activity

Launch

Launch the activity with a number talk: 15 + 5. Post the problem on chart paper or the board and provide students with think time. Invite students to share answers and then justify how they arrived at their solutions. As students share, highlight and name decomposing strategies that use 10s and 1s. For instance, if students think about the two 5s as making a 10 and then add the remaining 10, be sure to ask, Where did the 5 come from? Point out that the students *decomposed,* or took apart, 15 into 10 and 5 and then made a new 10. Ask, Why was it helpful to make a new 10? These discussions focus attention on the usefulness of thinking in 10s and 1s when joining and separating.

Show students a stick of 15 cubes and a stick of 5 cubes each. Ask, How could we show with unifix cubes what you did in your mind? Invite students to share ideas, and then you can model those ideas with the cubes for all students to see. You might, for instance, show taking 5 cubes from the stick of 15 and joining it with the stick of 5 cubes to make two 10s, which we call 20. Tell students that decomposing numbers into 10s and 1s can be a useful strategy for joining numbers, something they can try out in today's problems, which are all about recess.

Explore

Provide partners with the Play Ball! sheet and snap cubes, and read the task aloud. This is just the first of four tasks in the Recess Problem Set, and students can work through them at their own pace. Invite students to invent ways of solving the task and representing their solutions on paper. Encourage students to use the snap cubes to model the situations and justify their solutions. If students use snap cubes, ask them to draw on their paper what they did with those cubes and use labels to make their thinking clear. Partners consider the following questions:

- What is happening in this story?
- How could you model the action using objects, pictures, or numbers?
- How could 10s and 1s help you solve?
- How can you represent your solution on paper?

As you confer with students, support them in describing what they are doing with the numbers, particularly how they are using place value, or 10s and 1s, to decompose and recompose the numbers.

Discuss

Discuss and highlight both direct modeling strategies and those that use place value. Create a chart to show different ways that students solved the problems. Begin by discussing the Play Ball! task, but be sure to discuss other tasks that students tried. Discuss the following questions:

- What strategies did you develop for modeling and solving these problems?
- How did you show your strategies with cubes? How can the cubes prove that what you did made sense?
- How did you use 10s and 1s to help you?
- What did you notice about the numbers that helped you make new 10s?
- What strategies that we used made sense to you? Which strategies do you want to try in the future? Why?

Name the strategies that students share, both with mathematical language and by naming them for the students who shared them. Make connections between strategies that students used on different problems to show how using 10s and 1s can look different in different tasks but is still fundamentally the same idea.

Extend

Invite students to write their own recess problems to model and solve. Discuss the kinds of situations that come up on your playground or during recess time that involve joining (or separating). These might include numbers of kids moving between spaces of activities, scorekeeping in different games, counts of materials used on the playground, and the many things students might bring out with them, such as lunchboxes, jackets, or water bottles. Encourage students to make these problems represent their own recess and involve numbers that are realistic in their own experience. You might decide to write a shared task as a class, or ask each partnership to design their own. Depending on students' independence with writing, they may want to illustrate the situation rather than use written language to record it.

Discuss how students thought about writing the tasks and what strategies they used for solving. Make connections back to the chart of strategies created in the initial discussion and add to it when new strategies emerge. Continue to focus on the usefulness of 10s and 1s for solving problems.

Look-Fors

- **Are students solving symbolically (or mentally) and then justifying with manipulatives, or solving with manipulatives and then recording symbolically?** We encourage students to solve problems in multiple ways and to connect different forms of their solutions. For instance, in the number talk, students solve the problem mentally, but then you represent that thinking in writing, with numbers, symbols, images, and arrows. You follow this by asking how the cubes could be used to represent that thinking. Layering on these multiple representations of the same strategy builds connections between ideas, and the flexibility to use them in new ways. As you circulate, look for the entry points that students are using, their first strategies, and where they are actually solving the problem, and distinguish these from the representations they may layer on afterward. The ways that students are solving the problems tell you which ideas they feel most comfortable with, and the representations they layer on show the connections they are working to make.

- **Do you notice any students just counting on or counting all?** Counting strategies have been very useful for students and are encouraged. Students are still working on extending their counting of objects into larger and larger collections. It makes sense that they may use counting strategies in these tasks. However, as the numbers get larger, counting strategies, either counting all of the objects or counting on from the starting value, can become inefficient and prone to errors. While not taking away this strategy, ask, Is there a way you could solve this problem without counting each object? Could you model the situation in some way? Some students may be simply moving forward with counting without considering their options. Others may need to count the objects before being ready to brainstorm other ideas. You could also ask, How did you solve the number talk? If students used a strategy other than counting, you could invite them to try it in their current task.

- **Are students attending to place value?** Are students interpreting two-digit numbers as representing 10s and 1s? Be especially attentive to any strategies which emerge that decompose the numbers involved around place value. Some students will not fully appreciate that this is what they are doing without you drawing attention to it and naming their thinking. Students may simply treat the larger numbers as digits, saying things like, "I took the 3 and the 7 and made 10." When you notice this kind of explanation, ask questions and revoice to promote the idea of decomposing to use 10s and 1s. You might ask, "Where did the 3 and the 7 come from?" and you might revoice to say something like, "So you decomposed the 23 into 20 and 3?" Be sure to connect these numerical ideas back to concrete objects so that students know why they work. Ask students to show what it looks like, for instance, to have decomposed 23 into 20 and 3, and then how they used the numbers, now in the form of cubes, to recompose. Conceptually, we want students to appreciate that taking apart the digits actually involves separating the quantity into smaller groups.

Reflect

How can you use 10s and 1s to help you join numbers?

Reference

Carpenter, T. P., Fennema, E., Franke, M. L., Levi, L., & Empson, S. B. (2015). *Children's mathematics: Cognitively guided instruction*. Portsmouth, NH: Heinemann.

 **Play Ball!**

There are 22 balls in the yard for recess.

The PE teacher brings out 8 new balls.

How many balls are in the yard now?

 **Kids on the Playground**

There were 23 kids on the playground.

A new class of 27 kids joins them.

How many kids are on the playground now?

 **A Pile of Jackets**

There are 39 jackets in a pile at the start of recess.

11 kids take off their jackets and throw them in the pile.

How many jackets are in the pile now?

 **Playing Soccer**

18 kids are playing soccer at recess.

Then 13 more kids decide to play, too.

How many kids are playing soccer now?

Playing with Place Value across Problem Types

Snapshot

Students develop strategies using place value for solving joining and separating situations of a variety of types.

Connection to CCSS
1.NBT.4, 1.NBT.6, 1.NBT.2, 1.OA.8

Agenda

Activity	Time	Description/Prompt	Materials
Launch	10–15 min	Facilitate a number talk: 18 + __ = 30. Give students think time and then invite them to share and explain their solutions. Ask questions that support students in explaining how they imagined their solutions so that you can accurately represent them on paper or the board. Draw attention to strategies using 10s and 1s. Remind students of similar strategies they developed in previous activities, and introduce the tasks.	Optional: chart paper and markers
Play	30 min	Provide partners access to the Playing with Problem Types Card Deck and tools for modeling these tasks. Partners choose a problem to start with; for each problem, they work to interpret and model the story. Students develop strategies for solving, with a focus on using place value. Partners record their work on paper so that others can follow their thinking.	• Playing with Problem Types Card Deck, cut into individual cards • Make available: snap cubes and rekenreks • Optional: glue sticks or tape
Discuss	15 min	Discuss some of the problems students solved, how they interpreted and modeled the story, and the strategies they developed, particularly those that used 10s and 1s.	

Activity	Time	Description/Prompt	Materials
Extend	30+ min	Give students an open-ended problem, such as: We have a jar with 30 marbles. Some are red and some are blue. How many red and how many blue marbles could there be? Invite students to solve this problem repeatedly and look for patterns in their solutions.	• Make available: snap cubes and rekenreks • Optional: chart paper and markers

To the Teacher

This activity is designed to parallel the Play activity in Big Idea 3, and we draw on the same set of problem types described in that lesson: join—result unknown, join—change unknown, join—start unknown, separate—result unknown, separate—change unknown, and separate—start unknown. It is critical that students have the opportunity to engage with joining and separating problems of different types at the same time, rather than have these taught individually or in some artificial order. Part of the challenge for students is in interpreting the meaning of differently structured stories and figuring out ways to model those actions, as they did in Big Idea 3. As the numbers get larger, students similarly need to revisit these different problem types, grapple with interpreting and modeling them, and now incorporate ways of thinking about 10s and 1s.

Because both strategies and ideas about place value develop over time, this Play activity can be extended over weeks. Each day begin with a number talk, as we do here, that highlights changing problem types and encourages thinking in 10s and 1s. In the number talk in the launch, we use a sum that is a multiple of 10 to promote decomposing 10s. We then provide a deck of problem cards that can serve as a starting point for students' exploration of using place value to join and separate. We encourage you to create your own that include your students' names, situations that interest them and that they encounter in your school each day, and numbers that will challenge them to stretch and to develop their strategies.

In the extension, we offer a new problem type with multiple solutions, which students can solve repeatedly and look for patterns. In this type of problem, only one value is given, in this case the sum of red and blue marbles. Students can decompose this sum into different combinations of red and blue marbles and consider how making 10 helps them find viable solutions, such as 10 red and 20 blue, 11 red and 19 blue, 12 red and 18 blue, and so on. We encourage you to make this type of problem part of your problem set over time as well.

Activity

Launch

Launch the activity with a number talk: 18 + ___ = 30. Post this problem on a whiteboard or chart paper. Give students think time, and then invite them to share answers and explain their solutions. Chart students' strategies in ways that reflect their thinking. Ask, How could we see what you did? What picture could I draw to represent your thinking? Some students may think with counting, which can be represented on hands, while others may think with cubes; some strategies could be represented on a number line. Draw attention to strategies that use 10s and 1s or decompose using place value. For any of these strategies that students share, you might ask, How did thinking about 10s and 1s help you with this task?

Remind students of the strategies they developed in the Visualize activity (or in previous days of this activity) by referring back to the charts you created in the discussion. Introduce students to the problem deck; depending on the independence of students' reading, you may want to read a small number of these tasks aloud and invite partners to discuss which one they want to begin with.

Play

Provide students with access to the Playing with Problem Types Card Deck for students to explore and develop strategies for solving. Make available manipulatives for modeling solutions, such as snap cubes and 100-bead rekenreks. For each problem, partners develop a way to understand and represent the task and find a solution that they can show and explain. Students use the following questions as a guide:

- What is happening in this story? What do you need to figure out?
- How can you show what is happening in this story with objects or numbers?
- What strategies can you use to solve the problem?
- How can you show your thinking on paper? What pictures can you draw for others to understand what you did?
- How can you label your work to make your thinking clear?

For each problem that partners solve, they record their strategies on paper so that others can understand what they did. You may want to ask students to glue or tape down the problem card they solved onto their paper or in their notebooks to make it clear what problem they were working on.

Discuss

Students will have solved different problems at their own pace. In this discussion, you are likely to invite some students to share solutions to problems that others have not yet tackled. This is entirely appropriate, as the focus is on the thinking involved and developing strategies that use 10s and 1s. However, you may need to slow down to read aloud the problem being discussed so that everyone has a chance to think about the story before you share a particular group's solution. Consider choosing intentionally which groups will share, in order to highlight interesting, new, or useful strategies or challenges students faced that the class could grapple with together.

For each solution you discuss, ask the following questions:

- How did you understand what was happening in the story?
- How did you represent the problem? (Highlight strategies for modeling and for using place value to solve.)
- What strategies did you develop for solving the problem?
- Did you try any strategies that you saw others using (in the Visualize activity or on a previous day of this activity)?
- How did you use 10s and 1s to help you solve?

Throughout the discussion, focus on the ways that 10s and 1s, as numbers and as represented with cubes or other tools, helped students think about solving the tasks.

Extend

Present students with an open-ended task for which multiple solutions exist, such as the following:

We have a jar with 30 marbles. Some are red and some are blue. How many red and how many blue marbles could there be?

Be sure that students understand that we do not know how many blue marbles or how many red marbles there are and that multiple solutions exist. Invite students to find as many different ways of solving this problem as they can. Students then explore the question, What patterns can you see in your solutions? You might create a chart with different student solutions in a table, with Red Marbles, Blue Marbles, and Total Marbles columns, to help students look for patterns. Connect the patterns students see back to place value and making 10s.

Look-Fors

- **Are students accurately interpreting and modeling the stories?** In Big Idea 3, students worked with problems of the same types as the ones in this activity, but with smaller numbers. The change in the size of the numbers can pose conceptual challenges, making students feel stuck, even when they are familiar with these types of problems. Pay attention for students who might be struggling to make sense of the tasks or who feel overwhelmed before entering the tasks. Encourage students to return to the tools they developed in the past, such as acting out the story, modeling the action with objects, or retelling the story in their own words. You might read the story sentence by sentence to support making sense incrementally and ask, How could we model this part of the story? For problem types where the start or change is unknown, students will not be able to model the problem step-by-step but will instead need to think through how they imagine the action to model it. Talking it over among partners, with your presence to ask probing questions, can help them interpret the task and develop a way to enter it.

- **Are students decomposing numbers into 10s and 1s?** One key component of using place value strategically is thinking of any two-digit number as being composed of 10s and 1s. For instance, it can be useful to think of 16 as 10 and 6, or one 10 and six 1s, when joining it to a number like 22, which can similarly be thought of as 20 and 2. Decomposing into 10s and 1s enables students to use the commutative property (though they will not yet know that term for it) to change the order in which they combine the numbers. This decomposition reflects an understanding of our place value system that goes beyond seeing numbers as a continuous stream of counting. Place value is a way of organizing numbers, and that system can help us organize our thinking about them. Look for evidence that students are thinking about numbers as groups of 10s and 1s and ask them questions about the number they are working with to help them think about its meaning, such as, What do you know about 16? How can you model it? How can you use your model to help you join these numbers? Even if students have not yet figured out how to use place value, you can bring these ideas back to the whole-class discussion and invite others to think about how to do so.

- **Are students using trains of 10 to compose or decompose 10?** Another key component of using place value is understanding how a 10 is composed and how to use that understanding to solve problems. Students might consider a number like 40 and use their understanding of the trains of 10 (9 and 1, 8 and 2, 7 and 3, 6 and 4, and 5 and 5) to help them think about how to take a value such as 16 away from it. Alternatively, they might ask a question like, What joins with 16 to make 40? These two ways of thinking about the problem can both be supported by using the ideas that $6 + 4 = 10$ and that 40 is four 10s. Look for evidence that students are thinking about how to compose or decompose 10 using these pairs to help them as they solve problems. Ask questions to support students in developing explanations for their thinking. Some students may use their knowledge of making 10 but not yet have language for explaining. You can support their developing reasoning by pressing them to explain with words and to use objects to illustrate their thinking. Be sure to ask questions such as, Where did the 6 come from? Or, How did you know that you needed a 4? These questions will help students justify their thinking, both to themselves and to others.

Reflect

How can decomposing 10s and 1s help you solve problems?

Reference

Carpenter, T. P., Fennema, E., Franke, M. L., Levi, L., & Empson, S. B. (2015). *Children's mathematics: Cognitively guided instruction*. Portsmouth, NH: Heinemann.

Playing with Problem Types Card Deck 1

Kickball

16 first graders want to play kickball.
22 second graders want to play, too.
How many kids want to play kickball?

Time to Go Home

After school, 40 kids were climbing on the playground.
Then 16 kids went home.
How many kids are still on the playground?

Picking Flowers

Hugo picked 13 flowers at recess.
After school, he picked some more flowers.
Now he has 20 flowers.
How many flowers did Hugo pick after school?

Water Bottles

A pile of 20 water bottles was on a picnic table.
Then some kids grabbed their bottles and went inside.
Now there are 9 water bottles on the table.
How many water bottles did kids take inside?

Jump! Jump!

Some kids were jumping rope.
Then 21 more kids started jumping rope.
Now 40 kids are jumping.
How many kids were jumping rope at the start?

Backpacks Everywhere

Backpacks were all over the playground.
Then 24 kids took their backpacks home.
Now there are 25 backpacks left.
How many backpacks were on the playground at the beginning?

Basketball

12 student want to play basketball.
13 students join them.
How many students want to play basketball?

Morning Play

Before school, 23 students are on the playground.
Some of the students left for choir practice.
Now there are 19 students on the playground.
How many students went to choir practice?

Kickball

Some kids were playing kickball.
13 more came to play too.
Now there are 37 kids playing kickball.
How many kids were playing at the start?

Seagulls

There is a flock of 38 seagulls on the playground.
17 seagulls fly away.
How many seagulls are on the playground now?

Trees

There were 46 trees on the playground.
Some were sick and had to be removed.
Now there are 33 trees on the playground.
How many trees were removed?

Lunch Tables

There were a lot of lunch tables on the playground.
28 were taken away.
There are 31 lunch tables on the playground now.
How many lunch tables were on the playground at the start?

Making a Dollar

Snapshot

Students use coins to explore composing and decomposing with 100, connecting addition, skip counting, and place value.

> **Connection to CCSS**
> 1.NBT.2, 1.NBT.1, 1.NBT.4, 1.NBT.5

Agenda

Activity	Time	Description/Prompt	Materials
Launch	10–15 min	If not all of your students are familiar with coins, introduce them and make a chart of their names and values. Pose the task: If I have 100 cents in my pocket, what coins might I have? Give students a chance to revoice the task to ensure they understand the constraints. Invite students to discuss possible strategies or tools that could help them solve.	• Quarters, dimes, nickels, and pennies, to display and pass around • Optional: chart and markers
Explore	30+ min	Provide students with tools for modeling the task. Partners work together to develop strategies and find solutions. Students develop ideas about ways to combine and exchange coins and consider how to record and organize their thinking.	Make available: hundred chart, plastic coins, snap cubes, rekenreks, and base 10 blocks

Activity	Time	Description/Prompt	Materials
Discuss	15+ min	Discuss the ways that students thought about solving the problems, the strategies they developed, the tools they used to model, and what ways they found useful. Accumulate and organize the solutions that students have found and discuss patterns, including how to modify one solution to find another.	• Chart and markers • Coins, to display
Extend	45+ min	Explore how different constraints on the task change the solutions, such as by asking what the fewest or greatest number of coins might be for making 100 cents. Brainstorm additional questions for the class to investigate.	Make available: hundred chart, plastic coins, snap cubes, rekenreks, and base 10 blocks

To the Teacher

As with the open-ended task we suggested in the extension to the Play activity, this activity focuses on encouraging students to develop many ways of solving a single problem. Here, we use money as a framework for decomposing the number 100 using place value as a tool. Our US coinage system draws on place value, particularly with pennies representing 1s and dimes representing 10s. Nickels can be used to think of 5 both as half of a 10 and as five 1s. Quarters are trickier, drawing on all of these ideas, and can be thought of in many different and useful ways, including as twenty-five 1s, two 10s and five 1s, two-and-a-half 10s, a quarter of 100, and half of 50. Students are unlikely to have all of the conceptions of coins before beginning, but working with coins to solve this task can support students in developing ways of thinking flexibly about 10s and 1s and the relationships between place value and coins.

As students develop ideas about what coins could be used and in what quantities to make 100 cents, they will need some system for keeping track of their thinking and the solutions they have found. Encourage students to think about how they would like to organize their thinking. A table is a natural choice for adults, but many students will not think of this tool immediately. Rather than telling students to make a table, look for early examples of table-like organization, such as lists, and point out why these are useful. You may find that the opportunity to develop a class table presents itself in the discussion, and this would be an authentic way for tables to emerge from a need the class has. Tables are far more intuitive when they meet a need than when they are imposed.

This investigation can continue across multiple days, as you discuss some of what students have found, the strategies they developed, and the tools they used each day. On subsequent days, you can send students back to find new solutions, try out different tools, or build on the thinking they heard from their peers. If you have started a class table, you can add to it each day. At some point, this work can spur questions about whether there is an end to the number of solutions and how you will know when you've found them all.

Activity

Launch

If you don't think all of your students are familiar with the coins, launch the activity by introducing the quarter, dime, nickel, and penny. Name each one and chart its value in cents. Be sure that students know the term *cent* as the value of a penny and that 100 cents make a dollar. You'll want to give students a chance to see and touch each different coin.

Then pose the task, recorded on a chart or board:

If I have 100 cents in my pocket, what coins might I have?

Ask, What is this problem asking? Give students a chance to turn and talk with a partner to restate the task and what it is asking students to do. Invite students to revoice the task and ensure that everyone understands the constraints, particularly that we know how many cents but not what coins.

Ask, What could you try to solve this problem? Give students a chance to turn and talk about possible strategies. You might invite some students to share initial ideas, but be sure that no one gives a full solution at this stage. Students might suggest materials they could use to help them think, such as coins, cubes, or rekenreks.

Explore

Make available tools for modeling the task, including hundred charts, plastic coins, snap cubes, rekenreks, and base 10 blocks. Partners work to find solutions to the question, If I have 100 cents in my pocket, what coins might I have? Students use the following questions to guide their investigation:

- How can you make a dollar with coins?
- How can you make 100 cents in different ways?
- How can you use the values of the coins to build 100 cents?
- What combinations of coins are useful? Why?
- How can you record your solutions on paper?
- How can you find a new solution using the solutions you have already found?
- What patterns do you notice that can help you?

Students record all their ways on paper, showing how they arrived at their answers and organizing their thinking so that they can track and build on their solutions. Encourage students to consider how they can organize their thinking by asking questions as you circulate, such as, How can you organize what you've found so that you can keep track of your solutions?

Discuss

As a class, discuss the following questions:

- How did you think about solving this problem?
- How did you model it? What tools did you try?
- What tools or strategies were helpful (or not)?
- What mistakes did you make, and what did you learn from them?
- What solutions didn't work? Why?
- How did place value (or 10s and 1s) help you think about making 100 cents? (Highlight different ways of thinking about the task, such as adding up to 100 or decomposing 100.)
- How did you record your thinking?
- How can we as a class record the different solutions that we found? How could we organize the findings? (Consider developing a table to show the different ways students found to make 100 cents.)
- What patterns did you notice? How did you use them?

Be sure to have coins available to show the different solutions that students share. The visual representation of these solutions can help students focus not just on the numbers but on what those numbers represent.

This discussion can be a natural place to develop tables to organize information, and if that opportunity presents itself, seize it. Some partnerships might create something table-like as a tool, and these should be shared and built on for class organization. Tables can help students notice structures and patterns and might spur further investigation.

Extend

Invite students to continue their investigation by exploring different kinds of constraints on the task, such as those in the following questions:

- What is the fewest number of coins you can use to make 100 cents? Why?
- What is the greatest number of coins you can use to make 100 cents? Why?
- If you have to use at least one of each coin, then what is the smallest number of coins you can use to make 100 cents? What is the greatest?

Students might want to propose their own constraints or variations for further investigation. Those questions that come from students are often the best to pursue because they are based on authentic curiosity. You might offer the preceding questions as a starting point for brainstorming additional questions for the class or partnerships to choose from to deepen their thinking about making 100. Make available the tools students found useful in the initial activity and, afterwards, discuss how students thought about approaching the questions they posed and how their strategies changed.

Look-Fors

- **Are students struggling to manage all the different kinds of coins?** Because the goal of this investigation is to support students in using place value to solve joining and separating situations, it is entirely appropriate to encourage students to focus on the two coins that are closet to this goal: pennies and dimes. There are many combinations of pennies and dimes possible for solving this task, and these two coins, when used to compose 100, encourage students to think about making exchanges. For instance, if students make 100 cents with 10 dimes, they might then create a new solution by exchanging one of those dimes for 10 pennies. These exchanges reinforce the idea that 10 can be a unit (as in a dime) or a collection of 10 ones (as in 10 pennies). Students can focus on these two coins, and when they have built greater comfort, they can consider what new possibilities exist when they use nickels or quarters. The open-endedness of this task is intended to provide a low floor and high ceiling for students, not to overwhelm. Encourage students to think about how to simplify the task first and then steadily reintroduce complexity.

- **Are students using skip counting to count coins?** Coin counting is an authentic context for skip counting, but it can be conceptually challenging on at least two fronts. First, each coin is a single object, just like each snap cube or rekenreks bead, but for nickels, dimes, and quarters, those individual objects represent multiple cents. It can be conceptually challenging to touch one coin and simultaneously count "10." This can take practice and is in itself an abstraction. Students will need time to count coins, especially collections of the same coin, to build the connection between the object and what it represents. The second major challenge comes in switching between skip-counting intervals. In counting a collection of mixed coins, it makes sense to count all the coins of one kind and then switch to another, but that switch means that at some point in the counting stream, students have to change from skip-counts by, for example, 10s, to skip-counting by 5s or counting by 1s. Encourage students to slow down at these transition points and think through what comes next. You might simply pause students and notice aloud, "You've counted all your dimes and you have 70 cents. What comes next?" The hundred chart can be a great tool for making this transition visible and manageable.

- **Are students modifying solutions they have found to create new solutions, or starting from scratch each time?** You may see that students create a few solutions relatively quickly using what they already know about coins, and that each of these solutions is independent of the others. For instance, students might immediately say that they can make 100 cents with 100 pennies or 10 dimes or four quarters. Creating each solution from scratch means rethinking the coins and the number 100 over and over, something that will certainly help students build ideas about 100 and place value. However, there are even more ideas to be explored if students consider how to modify one solution to create a new one. Look for students making exchanges or thinking about equivalence to find new solutions. For instance, a student might have determined that four quarters can be used to make 100 cents, and then use their understanding of 25 to exchange one quarter for 25 pennies, or two dimes and a nickel, five nickels, or another combination of coins. This kind of systematic thinking uses the structure of place value and coins to generate solutions, a key mathematical practice, and one that attends to the relationships between numbers and values. Be sure to highlight that these are useful ways to approach the task if you see groups thinking in this way, and invite those who are to share their strategy in the discussion.

Reflect

What do you know about 100?

Using Units to Measure Our World

As I mentioned in Big Idea 1, mathematician Paul Lockhart considers the beauty of mathematics to be an "amazing adventure." All that is needed is imagination and curiosity. Lockhart has written an entire book about the subject of this particular big idea—measurement—and points out that all measurement is an approximation:

> The thing is, physical reality is a disaster. It's way too complicated, and nothing is at all what it appears to be. Objects expand and contract with temperature, atoms fly on and off. . . . Nothing can be truly measured. A blade of grass has no actual length. Any measurement made in this universe is necessarily a rough approximation. . . . The smallest speck is not a point and the thinnest wire is not a line. (2012, p. 1)

We may measure something and think that our measurement is precise, but it will never be a precise number but always an approximation. We enter this big idea in the spirit of approximation, hoping that our lessons will provide a good opportunity to teach students the value of being approximate and of estimating. Many children and adults think it is somehow wrong to estimate and that mathematics is all about precision. I have known students who, when asked to estimate, calculate precisely and round their answer. I think students do this because they do not believe that estimation is a valid mathematical act, and they may also struggle with number sense, which helps people estimate. A UK report determined estimation to be the most used mathematics in people's jobs and lives (Cockcroft, 1982). Here we help students learn that two important mathematical acts are estimation and approximation, and we bring these to students in the context of measuring. To help students measure we suggest objects in the world, which gives students an opportunity to exercise important brain areas that are different from those used when the mathematics is on paper. We think students enjoy interacting with the world as they develop ways of measuring.

In our Visualize activity, we help students learn the value of measuring with nonstandard units. Students first consider two objects in the classroom that cannot be put side by side, and they are asked, which is longer? Students then develop their own tools to measure. These skills are likely to be used in situations when a standard measure would not work, and students are encouraged to find a nonstandard measure of their own. Next the students choose their own objects and decide which is longer, justifying with their chosen measurement tools.

In the Play activity, students are invited to estimate visually, using important mathematical brain pathways. Students look at three joined snap cubes and work out what objects in the classroom might be a similar length. This is a moment to value approximation and to discuss this idea with students. As the lesson continues, students are given a set length of cubes to measure with and they go around the classroom collecting objects to make a visual display of the items that meet their measurement. Students can make a display of their collected data and the class can review each other's displays in a gallery walk.

In our Investigation activity, we bring in one of my favorite mathematics manipulatives. Students are now invited to measure classroom objects using the many choices of Cuisenaire rods. Students then work on some important data science options as they prepare displays of their data. Students can develop and label data visualizations, such as charts. As students develop charts and record their data they can feel ownership of their work and what they have created, which is always an important part of mathematics.

Jo Boaler

References

Cockcroft, W. (1982). *Mathematics counts*. London: Crown.

Lockhart, P. (2012). *Measurement*. Cambridge, MA: Harvard University Press.

Which One Is Longer?

Snapshot

Students develop strategies involving iterating nonstandard units for comparing the lengths of two objects without direct comparison.

Connection to CCSS 1.MD.1, 1.MD.2	

Agenda

Activity	Time	Description/Prompt	Materials
Launch	10 min	Point out the two classroom objects you have selected and ask, Which one do you think is longer? Give students a chance to turn and talk and discuss their predictions and reasoning. Ask, How can we prove it? Give students another chance to turn and talk to develop ideas, and then tell students that developing strategies for comparing is their goal for this activity.	Two classroom objects that are similar in length but cannot be held side by side (See To the Teacher)
Explore	20–25 min	Partners develop strategies and ways to use different tools to determine which object is longer. Partners collect evidence to share with the class.	• Two classroom objects that are similar in length but cannot be held side by side • Make available: nonstandard tools for measuring, including square tiles, snap cubes, sticky notes, and any other classroom material students want to use

Activity	Time	Description/Prompt	Materials
Discuss	10–15 min	Discuss which object is longer using the evidence students collected. Compare the strategies students developed and draw attention to the iterating of units to quantify length. Name the *unit* as the length that gets repeated to measure. Discuss what made some units more precise or easier to work with than others.	
Extend	30+ min	Partners develop strategies for determining how much longer the longer object is than the shorter one. Students may try new units or tools for measuring. Discuss the strategies they developed, their experiences using new units, and how much longer one object is than the other.	• Two classroom objects that are similar in length but cannot be held side by side • Make available: nonstandard tools for measuring, including square tiles, snap cubes, sticky notes, and any other classroom material students want to use

To the Teacher

For this activity, you'll need to choose two objects or distances in your classroom that cannot be moved side by side, such as two bookcases, the lengths of two rooms, or two tables, for students to compare. The items do not need to be of the same type; you could ask students to compare the length of a bookcase to the length of a table. If students cannot hold the objects next to one another for direct comparison, they will need to develop some other way to compare the lengths of the objects, particularly iterating units. Students may use a variety of tools and manipulatives, including snap cubes, sticky notes, or parts of their own bodies, to serve as the unit. The goal is to promote the development of iterating units, whereby students take a particular length (for example, the length of a square tile) and repeat that length to measure a linear distance.

Although eventually it will become useful for students to use standard units, such as inches, feet, or centimeters, to measure the lengths of objects, we want to avoid standard units in this activity. Before students can develop a conceptual understanding of standard units, they need to know why such units are necessary or helpful. One key reason is that standard units support iterating same-size units, because they can be arrayed on a tool such as a meter stick or ruler. This activity opens up the concept of iterating and offers opportunities to discuss why units of the same size are needed to make meaning out of counting them. Even if your students have been exposed to standard units, focus attention on using practical objects by not putting out rulers or other standard measuring tools. If students ask for them, redirect by asking what else they could use instead, making sure that any other tools in the classroom are considered fair game for use as a unit. These might include school supplies (pencils, markers, or paper clips), manipulatives (snap cubes, square tiles, or base 10 rods), everyday objects (picture books, shoes, or floor tiles), or parts of students' bodies (hands, feet, or wingspan).

Activity

Launch

Launch the activity by pointing out the two objects you have selected in your space that are similar, but not the same, lengths and cannot be held side by side. Ask, Which one do you think is longer (or taller)? Why? Be sure to be clear about which dimensions are being compared. Give students a chance to turn and talk to a partner, and then discuss students' predictions and their reasoning. Some students may point out that it is hard to know, and if they do, ask why. Open up discussion about what makes it easier to compare the lengths of two objects.

Then, drawing on students' predictions about which object is longer, ask, How can we prove it? Give students a chance to turn and talk to a partner about the strategies they could use to find out which is longer. Instead of discussing these strategies as a class now, tell students that it is their job today to develop strategies for figuring out which object is longer and to provide evidence to the class.

Explore

Partners develop a plan for comparing the two objects using whatever tools make sense to them. Provide access to some materials, such as square tiles, snap cubes, sticky notes, and any other nonstandard measurement tool students might ask for. Students consider the following questions:

- Which object is longer?
- What tools or strategies could help you prove which object is longer?
- How can you tell how long an object is?
- How can you record your evidence to share with others?

As you circulate, ask questions about the strategies for comparison that students are developing, why they are choosing particular tools to support comparison, and how they might be using those objects.

Discuss

As a class, discuss the following questions using the evidence students collected:

- Which object is longer (or taller)?
- How did you decide?
- What tools did you use? How did you use them?
- What challenges did you face, and how did you solve them?
- How confident are you in your measurements? Why?
- How did you decide how long an object was when your units didn't fit precisely?
- How could we have been more precise or accurate?

During this discussion, name that when we use one size of object over and over to measure, that object is a *unit*. Focus on highlighting how students iterated units (or tried to) and the issues that arose when they tried to do this. Use the language of *unit* to discuss the tools they used to measure, as in, "So Marquis and Delilah used sticky notes as their unit. Did anyone use a different unit to measure?" Be sure to dig into the challenges students faced in using the units they selected, pointing out where precision was difficult and why. For instance, students may have been more precise using snap cubes—because they connect end to end—than they would be using pencils, which have a tendency to roll.

Extend

Invite students to try to quantify in some way the lengths of the two objects to decide: How much longer is one than the other? Students may want to measure in entirely new ways after seeing new ideas in the discussion. Make available the same wide variety of classroom objects and manipulatives from the first part of the activity, plus anything new that students might request. Encourage students to think about the units they might use to help them decide how long the objects are. This is both a new measurement task and a new type of comparison problem, which students often find challenging. Before starting the task, be sure that students can visualize and revoice what it means to ask "how much longer" one object is than another. After students have had a chance to explore and gather evidence, discuss the strategies they used, their experience using new tools if they did so, and how much longer the one object is than the other.

Look-Fors

- **Are students using the same unit of measure, or are students mixing units?** For a measurement to be quantifiable and comparable, students need to be using the same unit for the full length of both objects. For instance, we can compare the length of a bookcase that is 86 snap cubes long to a table that is 67 snap cubes long because the same unit is used consistently. However, students might mix units within an object (such as markers and crayons), use inconsistent units (such as pencils of varying lengths), or use different units for the two objects (such as sticky notes for a bookcase and snap cubes for a table). Any inconsistency will make comparing the lengths of the two objects unreliable. It might be tempting to tell students this directly, to avoid all the time spent on measuring and later remeasuring (as well as the accompanying frustration). However, these challenges are exactly what we hope will emerge from this activity. Rather than avoiding these problems, focus on them when they arise, and discuss what makes it hard to compare or describe the lengths of the objects using the units they have chosen. Support students in thinking through new ways to address the issues they encountered and developing new strategies. Name the challenge that students faced and then ask, What else could you try to make that easier?

- **How precise are students being in measuring?** One of the reasons we use standard units is that we have tools, such as rulers and tape measures, that make it easy to iterate those units with some precision. Working with nonstandard units, particularly those that do not connect, means that students will be grappling with how to line up the units. Look for the ways that students think about this challenge. Are they attempting to arrange their units end to end, or are they leaving gaps? If there are gaps, you may want to ask students about those spaces by saying something like, "I notice that there is this empty space here between your sticky notes. How are you counting that part of the length?" You may also see units arrayed in nonlinear ways, such as a snaking line of pencils or an angled line of square tiles. You may want to take photos of these attempts to discuss as a class, particularly if different student groups use the same tools and arrive at different lengths for the same object. These photos could help the class discuss why one group found the length to be, for example, 12 pencils, while another found it to be much more or less.

- **How are students dealing with partial units?** Often students will find that an object is some number of units long and "a little more." This will be most frequent with students using longer units, such as markers, pencils, or their own hands. How do students deal with that "little more"? Some students may decide simply to describe the length as "a little more than" 8 pencils long, for instance, while others may try to quantify this leftover distance. Students might refer to any partial length as "half" regardless of its actual distance or use a different unit that fills the space, as in, "This table is 8 pencils and one eraser long." Discuss as a class whether those leftover distances should be counted at all and if so, how. As you circulate, ask students questions about these leftover distances, such as, What about this extra part at the end? Does it count? How could you measure it? How long is it? These questions encourage students to reason about precision and partial units.

Reflect

How can you tell how long something is?

Measuring Our World in Cubes

Snapshot

Students play with estimating and measuring lengths using snap cubes, collecting items of similar lengths from around the room to create a measurement gallery.

Connection to CCSS
1.MD.2, 1.MD.4

Agenda

Activity	Time	Description/Prompt	Materials
Launch	10 min	Show students a stick of three snap cubes and ask, What can you see in our classroom that you think might be 3 snap cubes long? Give students a chance to generate ideas and then have them retrieve one object from the classroom. Discuss how to prove whether their objects are 3 cubes long and then compare each to a stick of three cubes. Discuss how students made their estimates and the challenges they faced.	Stick of three snap cubes
Play	20–25 min	Assign each partner a length of cubes using the Measuring the World in Cubes Cards and provide cubes for making sticks of that length. Partners search for objects in the classroom that match the length they are assigned and then create a display of those objects.	• Measuring the World in Cubes Cards, a different card per partnership • Snap cubes • Optional: masking tape

Activity	Time	Description/Prompt	Materials
Discuss	15 min	Do a gallery walk of students' displays and test the objects' lengths. Discuss how students were thinking when they were searching for objects, what surprised them, and how they tested their ideas. Invite students to share observations and surprises from the gallery walk.	Students' displays of collected objects, snap cubes, and their Measuring the World in Cubes Card
Extend	5–10 min	As a regular routine, do a length estimation talk by showing students an object and inviting them to estimate how many cubes long it is. Students think, turn and talk to a partner, and then share their estimates and reasoning. Then build a stick of cubes together until the class can agree on the actual length of the object.	• Classroom object • Snap cubes • Optional: chart and markers

To the Teacher

Often when students are learning to measure, we focus attention on the process of measurement and ignore the role that estimation plays in understanding the meaning of units and length. In this Play activity, we bring these ideas together, encouraging students to connect quantities, number, measurement, units, and length through repeated estimation and measurement. We use snap cubes as the shared unit so that students can compare the objects that different groups collect, with the idea that objects 12 cubes long should be longer than those that are 10 cubes long. As students work to collect objects from around the room that are approximately the length they have been assigned, they will need to estimate with their eyes, noticing objects that might be the right length, and then test them. These repeated experiences will build some fluency with estimating using particular lengths, just one step in a longer developmental process of connecting number, unit, and length to the real world.

In the extension, we offer a routine for continuing to develop students' visual estimation capacity. You can use these estimation talks as an initial activity before any other lesson to provide ongoing opportunities to think about length, unit, and quantity.

Activity

Launch

Launch the activity by showing students a stick of three snap cubes. Ask, What can you see in our classroom that you think might be 3 snap cubes long? Give students a chance to turn and talk to a partner about their ideas. Then invite partners to retrieve from the room one of the objects they thought of. Ask, How could we test whether the object you predicted is actually about 3 cubes long? Takes some ideas and then invite each partnership to test whether the object they selected is about 3 cubes long by directly comparing it to a string of three cubes.

Ask, How did you make your prediction? What made it challenging to find an object 3 cubes long? Discuss students' strategies for estimating and the challenges they faced, particularly that they could not hold the three-cube stick in their hands to see or feel how long it was or directly compare it to the objects in the room. Tell students that the more experience they have estimating and measuring length, the more accurate they will get.

Play

Assign each partnership a different length of snap cubes using the Measuring the World in Cubes Cards. Note that we have left a few blank if you'd like to make your own. Provide students with snap cubes to build the length they have been assigned. For longer lengths, students may want to use masking tape to ensure that their sticks don't break repeatedly as they work. Partners explore the question: What can you find in the classroom that is ___ cubes long? Partners work together to make predictions about the object that might be their designated length and then measure each. Partners collect the objects they find that are the appropriate length. If students choose objects that can't be moved (such as a piece of furniture or a poster on the wall), they might jot down the object's name or make a sketch of it on a sticky note that they can add to their growing collection.

Partners then make a visual display of the objects in their collection, like a museum exhibit, and use their Measuring the World in Cubes Card to label the display along with the stick of cubes they made to measure the objects.

Discuss

Do a gallery walk of students' displays and invite partners to test measuring the objects using the cubes in the display. As they walk, students think about and explore the question, How close are these objects to the length of cubes in the display? Give students time to play with the objects at multiple displays and make observations about the different stations.

Then, as a class, discuss the following questions:

- What were you looking for when you were searching for objects that matched your length?
- What surprised you?
- How did you test your ideas?
- How did you deal with things being not exactly ___ cubes long?
- What did you notice on your gallery walk? What surprised you?

Extend

As a regular routine similar to a number talk, do an estimation talk. Show students an object from your classroom, such as a marker, book, bin, or the height of a table. Ask, How long do you think this is in cubes? Give students a chance to look and think, then turn and talk to a partner about their estimate and reasoning. Ask, How many cubes long do you think it is? Record students' estimates on a board or chart and ask for their reasoning. Students may use previous objects as benchmarks (as in, "It looks like the same length as a pencil, and I know that is 8 cubes long") or visualize the cubes in their minds.

Then build a stick of cubes one by one to measure the object until the class can agree on a length. You can do one of these estimation talks each day to support students in building connections between quantity, number, length, and, if you select longer objects, place value. You can choose new objects or from among the objects in students' displays from this activity, asking students' whose object it is to not give away the answer. Students will enjoy having their objects chosen for this routine.

Look-Fors

- **Are students testing their objects?** This activity is intended to engage students in both estimating and checking the length of objects repeatedly. For instance, students may see an object from across the room that they estimate may be the length they are looking for, based either on the stick of cubes they have constructed or other objects they have already measured. Then the students need to test that object against a known length to determine whether it is, in fact, approximately the desired length. Look for students simply estimating what looks right and collecting objects without checking them in some way. Ask, How do you know that the object is about the right number of cubes long? What evidence can you gather for your idea? How could you check?

- **How close are students' estimates when they select objects? How precise are students being in selecting objects?** As students estimate with the same length repeatedly, one goal is that their estimates become increasingly precise as they develop a visual sense of their length. As they accumulate objects that are the desired length, they can mentally compare new objects to both the stick of cubes and all of the others they have measured; each object can become a benchmark or reference for other objects. As the number of objects measured climbs, so should the number of reference points students can draw on to think about the length of other objects. However, some students may be grabbing and checking objects indiscriminately, which will not support them in becoming more precise with their estimates. If you see students checking one object after another without an apparent plan, ask, How did you decide that this object might be the right length? Is there any way you can decide whether it is close to the right length before you measure it? Some students may want to collect all (or many) of the objects they test, making piles for those that are too short and too long, keeping their references handy so that they can estimate more accurately. You might ask students to step back, look at their stick of cubes or other reference items, and then look around the classroom, thinking aloud with you about objects they might test before they go off to do so.

Reflect

What helps you estimate the length of an object?

What can you find that is about 10 cubes long?	What can you find that is about 23 cubes long?
What can you find that is about 15 cubes long?	What can you find that is about 7 cubes long?
What can you find that is about 25 cubes long?	What can you find that is about 6 cubes long?

What can you find that is about 21 cubes long?	What can you find that is about 18 cubes long?
What can you find that is about 11 cubes long?	What can you find that is about 13 cubes long?
What can you find that is about 9 cubes long?	What can you find that is about 17 cubes long?

What can you find that is about 14 cubes long?	What can you find that is about 22 cubes long?
What can you find that is about ____ cubes long?	What can you find that is about ____ cubes long?
What can you find that is about ____ cubes long?	What can you find that is about ____ cubes long?

Measuring with Cuisenaire Rods

Snapshot

Students explore the role that units play by measuring a shared collection of classroom objects using different Cuisenaire rods as units.

Connection to CCSS
1.MD.2, 1.MD.4

Agenda

Activity	Time	Description/Prompt	Materials
Launch	10 min	Remind students that they have used cubes to measure, and show them the set of Cuisenaire rods as a new tool for measuring. Create with the class a set of 6–8 classroom objects to measure. Show students the table you have created for recording the lengths of the objects in the different rod units.	• Cuisenaire rods, to display • Collection of 6–8 objects to be measured, collected as a class • Chart set up to collect measurements (see To the Teacher)
Explore	30+ min	Partners choose objects from the class collection and measure the length of each using different Cuisenaire rods as units. For each object students measure in each unit they choose, partners add their measurement to the class chart.	• Collection of 6–8 objects to be measured, to be shared among the class • Cuisenaire rods, per partnership • Chart set up to collect measurements, and markers
Discuss	15 min	Discuss how students measured the objects, the challenges they faced, and how they dealt with units that did not fit. Discuss what happened when they measured the same object with a different unit. Examine the class table and look for patterns. Focus attention on the relationship between unit size and the number of units needed to measure an object.	• Class chart of measurements • Collection of 6–8 objects the class measures • Cuisenaire rods, to display

To the Teacher

For this activity, you will need to construct a collection of six to eight classroom objects that students can measure. You may want to select some of these in advance. It is useful when selecting the objects to be measured if at least some of them are items that you have multiple examples of, such as markers, notebooks, or chairs. This way there are enough "copies" of the objects in the collection to go around. However, it is also useful if the collection contains objects students are genuinely interested in measuring, and we encourage you to ask students to come up with at least some of the objects in your collection. Once you have determined which objects the class will measure, you will also need to specify what dimension students are measuring so that the lengths that different groups find are comparable.

In advance, set up a table on a chart or board that students can reach with rows for each object (6–8 rows) and columns for each unit students might use to measure it: white (1), red (2), light green (3), purple (4), yellow (5), dark green (6), black (7), brown (8), blue (9), and orange (10). As students find the lengths of each object using whatever units they select, ask them to contribute their findings to the chart. For instance, you might have a pencil on your chart that student groups determine is 5 light green rods long, or 4 purple rods long, or 3 yellow rods long, and each of these measurements will go in a cell of the table. As students work, they might use the table to find new ways of measuring the objects. For instance, with the example of the pencil, partners might see that the pencil has already been measured with light green, purple, and yellow rods, and they may ask instead, How many orange rods long is the pencil?

During this investigation and particularly in the closing discussion, we encourage you to press on two ideas. First, some units are more precise because when iterated they fit the length better, leaving little extra left over. The most precise measurements, then, can be found with the smallest unit, and, conversely, the longer units often leave the greatest length unaccounted for. Second, larger units need few iterations to measure the object, while smaller units require more. This inverse relationship is a key idea in measurement, and one that will come up when students move to standard units in later years.

Activity
Launch

Launch the activity by reminding students that in the Play activity, they used cubes to measure objects. Tell students that today they will try using different units. Show students a collection of Cuisenaire rods as tools for measuring a set of objects and

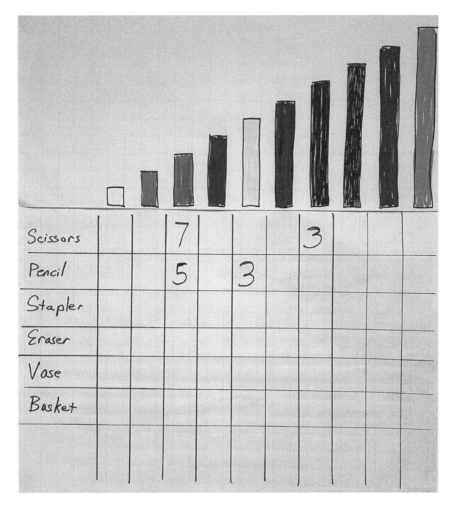

A chart showing student measurements using Cuisenaire rods

remind them that these rods come in different lengths. You may want to ask, What do you remember about Cuisenaire rods? Discussing this question can be a way to reconnect students with these tools if they have not used them for a while.

Create a set of six to eight classroom objects for students to measure during the Explore part of the activity. You may have selected some items in advance, but we encourage you to engage students in cocreating the collection. Tell students that they will be asking the question, How long is this object? Be clear about what dimension the class will be measuring on each object. They can use any of the Cuisenaire rods they choose to find the length of each object.

Show students the table you have set up to record their findings. Record the names of the objects to be measured in the left column. You may also want to make small sketches of the objects to remind students what the names represent. Encourage students to try out different rods for measuring the objects and show students where in the table their measurements will go depending on which rods they select.

Explore

Provide each partnership with a set of Cuisenaire rods and access to the objects to be measured. Partners work together to measure the length of each object using whatever Cuisenaire rods they choose. Students consider the following questions as they explore:

- Which Cuisenaire rods will you use to measure the object?
- How long is the object?
- What happens when you measure the same object with different rods?
- If the rods don't fit exactly, how can you describe how long the object is?

Encourage students to measure with different rods side by side, as shown in the image here. As you circulate, ask students which units give the most precise measurement for a given object. As students find the length of each object, ask them to contribute their measurements to the class table. Students might use the table as a way to choose the units they use so that they can contribute new measurements to the chart.

Discuss

After students have had a chance to make and contribute multiple measurements to the class chart, discuss the following questions as a class:

- How did you measure each object? Were some harder to measure than others? Why?
- What happened when you measured the same object with different units?
- Were some units better than others? Why?
- What challenges did you encounter?

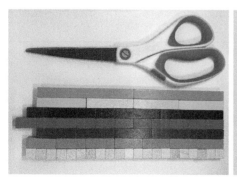

Cuisenaire measurements for scissors and a pencil

Look at the class table together and discuss the following questions:

- Why did we get different measurements for the same object?
- What patterns do you notice?

Focus attention on the pattern that with the smaller units, there are always more needed to describe the length than with larger units. If students notice this pattern, ask them to reason about why this might make sense. Students may want to get an object to show why smaller units require more iterations than larger units.

Look-Fors

- **How are students describing the lengths when the units don't fit precisely?** Students may naturally struggle to know how to name the length of an object when no whole number of units seems to accurately describe how long it is. This may be particularly true when students use a long unit (such as the brown, blue, or orange rods) to measure a relatively short object, such as a marker or eraser. Students may variously decide to use the whole number of units that fit (erring on the short side), use the whole number of units that are required to fully cover the length (erring on the long side), or use a fractional name (such as one and a half). Some students may, alternatively, decide that the unit cannot be used to measure that object. Ask students questions about why they decided to measure in the way they did and be sure to raise this issue in the discussion so that the class can debate what they think makes the most sense. This is not a decision with a clear and accurate answer given the conceptual tools available to first graders, and it is therefore a site for genuine mathematical argument. Embrace it!
- **Are students comparing the units and noticing that some work better than others for particular objects?** Particularly if students measure the same object with different units side by side, they may notice that a whole number of one unit might very closely match the length of a particular object, but that this is not true of all units. Use this as a place to discuss precision. Ask, Which unit is the most precise for describing the length of this object? How do you know? For instance, students might see that a pencil can be described as either 3 yellow rods or 8 red rods long, but that the yellow rods are a more precise description of the pencil's length because they begin at the eraser and end at the tip. You might ask students to indicate on the chart when they think they have a precise length, in order to open up discussion about patterns in precision.

- **Are students reasoning about why the *number* of units changes when they change units?** It is critical that students see that each time they measure with a new unit, they need to count those units and, in doing so, see that the number of units needed is likely to change. Before students get to the discussion and examine the class chart for patterns, it is useful if students begin to reason about why the number of units changes each time they use a different unit for the same object. When you see a group measure an object more than once with different rods, ask them to talk about what they found. Revoice their finding and ask questions about why the numbers are different, as in, "You said this pencil is 3 yellow rods long and that it is also 8 red rods long. Why are those numbers different? Why did it only take three yellow rods, but then take so many more red rods?" In these conversations as you circulate, students can begin to grapple with the meaning of changing units, and they will be better able to grapple with the larger pattern in the table.

Reflect

Why do units matter when measuring length?

BIG IDEA 10

Partitioning Shapes into Equal Parts

The composing and decomposing of geometric shapes is an important part of students' mathematical understanding. As students think and reason about the ways that smaller shapes combine to make bigger shapes and bigger shapes come apart as smaller shapes, they start to learn the composition of shapes. Researchers have studied how students learn through shape puzzles and found that they provide a perfect way for students to develop a deep spatial understanding (see, for example, Polly, Hill, & Vuljanic, 2015). The Common Core State Standards in Mathematics set out the need for students to build composite shapes from smaller shapes and break geometric shapes into smaller shapes. The Progressions of the Common Core (https://www.cgcs.org/cms/lib/DC00001581/Centricity/Domain/120/ccss_progression_sp_hs_2012_04_21_bis.pdf) explain that this type of building and separating helps students develop the ability to compose and decompose numbers.

As students play and learn with spatial puzzles, they will learn that shapes, just like numbers, can be understood flexibly. One of the activities I like to give people is what is known as the Marshmallow Challenge (https://www.tofflerassociates.com/vanishing-point/why-kindergarteners-always-win-the-marshmallow-challenge). In this challenge, people are asked to build the biggest freestanding tower they can that will support a marshmallow on the top. They have the following resources: 20 sticks of spaghetti, one yard of tape, one yard of string, and one marshmallow; they have a time constraint of 18 minutes. In Tom Wujec's TED talk (https://www.ted.com/talks/tom_wujec_build_a_tower_build_a_team?language=en), he reveals that when he has given this challenge to different groups, including graduate students and businesspeople, kindergarteners have been the winners. I have found similar results in my own teaching, with young children building higher and better towers than adults. The reason, it is believed, is that the young children experiment more. Whereas adults often plan what they think is a stable structure, the young children keep iterating with parts of the structure, and they build with more freedom and creativity. The set of activities in this big idea are designed to harness the abundant curiosity and creativity that first-grade students will bring to the activities, in the service of important learning.

In our Visualize activity, students are asked to cut rectangles in half so that they can see and celebrate all the different ways they see a half, and learn the definition of a half. Encourage students to discuss the ways that were easy to notice, challenging, or surprising, and of course, to celebrate any mistakes. This task will also give students opportunities to work on rotation and flipping. I meet many adults who say that they have trouble imagining rotations and do not have the "right" brain for it, but all brain areas grow and develop with practice, and this activity is a great opportunity for that.

In our Play activity, students are given tangram puzzle outlines to match with shapes. Students are asked to make shapes and will have more opportunities to rotate and flip objects and consider the relationships between large, medium, and small triangles. This lesson is all about composing shapes using tangram pieces and thinking about what the new shape might be a part of or equal to.

In our Investigate activity, we invite students to investigate the symmetry of different flags. We have chosen flags with particularly interesting shapes that we think students will be curious about. This lesson is an opportunity to value the different countries that students know or come from. Flags can generate many types of rich conversation. In our lesson, we ask students to think about halves, thirds, and fourths and the ways they arise in a real situation.

Jo Boaler

Reference

Polly, D., Hill, T., & Vuljanic, T. (2015). Students' experiences composing and decomposing two-dimensional shapes in first and second grade mathematics classrooms. In D. Polly (Ed.), *Cases on technology integration in mathematics education* (pp. 121–142). Hershey, PA: IGI Global.

Half and Half-of-a-Half

Snapshot

Students explore the meaning of half by partitioning rectangles and other shapes, and then extend their thinking to include fourths and half of a half.

Connection to CCSS
1.G.3

Agenda

Activity	Time	Description/Prompt	Materials
Launch	10 min	Show students a rectangle on the document camera and ask, How could we cut this rectangle in half? Invite one student to share, and discuss as a class whether the shape has been cut in half and how they know.	• How Could You Cut This Rectangle in Half? sheet, to display • Marker
Explore	20+ min	Partners work together to find as many ways as they can to cut the same rectangle in half. Students can extend their thinking about half by exploring ways of cutting other shapes in half.	• Cutting Rectangles sheet, one or more per partnership • Make available: scissors, rulers, and Cutting Shapes sheets
Discuss	10–15 min	Display students' ways of partitioning the rectangles, discuss which ways create halves, and develop a definition of *half*. Discuss which ways of cutting in half were easy to notice, challenging, surprising, or mistaken. Discuss the other shapes that students partitioned and what they tell us about half.	• Students' partitioned rectangles, displayed

Activity	Time	Description/Prompt	Materials
Extend	40+ min	Repeat this activity focused on partitioning shapes into fourths and building a shared definition of *fourths*. Discuss the relationship between halves and fourths and support students in understanding that a fourth is a half of a half.	• Cutting Rectangles sheet, one or more per partnership • Make available: scissors, rulers, and Cutting Shapes sheets

To the Teacher

In this activity, we create an opportunity for students to draw on their everyday knowledge of half to develop a more formal mathematical definition. First graders typically have had many experiences cutting things in half to share with a friend, partner, or sibling, though sometimes these experiences are imprecise, leading to disagreements about whose half is bigger. Here, we encourage a focus on conceptual precision, where students cut shapes and debate whether they have created halves to move toward understanding halves not just as two parts but as two *equal* parts. We begin with rectangles because they offer a familiar space to explore half in many different ways. Some students may want to apply their thinking to other shapes, so we have provided some additional shapes for exploration, which can challenge students to become even more precise with their thinking.

We highly encourage you to do the extension activity, which simply revisits the same structure to build an understand of fourths as they relate to halves. While students often have used the term *half* in their everyday lives, they may not have encountered *fourth* or may only be familiar with the idea from the term *quarter*. You may need to have more discussion as a class during the launch of this extension to support students in thinking about what *fourth* could mean. Also worth noting here are some linguistic challenges. The word *half* has no relationship to the number word *two*, and students are unlikely to get these confused, either as words or concepts. This fraction terrain shifts with *fourths*, which often sounds in speech very much like the number word *four*. Be cautious in introducing this new term to distinguish fourths from four; recording the word where students can see it and pressing students to be precise themselves will help.

Some of your students are likely to want to cut shapes physically, rather than just marking them with a pencil. Cutting or folding has real benefits for being able to see whether the two parts fit on top of each other. If you have a die-cut machine in your school, you might cut shapes for students to explore cutting and folding in half and fourths without the need for students to cut out the shapes from our sheets. When it comes to selecting shapes, those with symmetry present more opportunities to see how two regions are equal.

Activity

Launch

Launch the activity by showing students the How Could You Cut This Rectangle in Half? sheet on a document camera. Ask, How could we cut this rectangle in half? Give students a chance to turn and talk about ideas. Invite one student to come up and show how they could partition the rectangle into two halves.

Ask the class, Do you agree or disagree that this is half? Why? As a class, discuss what it means to be "half." The class may not yet agree on what makes half, and this is appropriate at this point. However, as you talk about half in the launch, point out any criteria about which students disagree or are not yet certain. For instance, students may say that the two parts, or halves, must be the same, but they may not yet be able to describe how they are the same.

Explore

Provide partners with the Cutting Rectangles sheet. Partners work together to find as many different ways to cut these rectangles in half as they can using their pencils to mark how they would cut each. Some students may want to cut out their rectangles and use folding or scissors to physically cut to find ways of partitioning into halves. Provide access to additional sheets, if needed; scissors; and rulers to make straight edges.

If students want to extend their thinking about what makes half, they can try to partition the figures on the Cutting Shapes sheets as well.

Discuss

Discuss and display these different ways that students have devised for cutting rectangles in half and develop a definition of *half* as something cut, divided, or partitioned into two equal parts. Students may need to explore what *equal* means in this context. We typically view equal with figures as equal area, but students have not yet developed ideas about area. They may simply say that the two parts are the same shape or that they take up the same space.

Then discuss the following questions:

- Which ways to cut the rectangle in half did you notice easily? Why?
- Which ways of cutting the rectangle in half were harder to notice? What did you try?
- Which ways to cut a rectangle in half did you try but found didn't work? How did you know it wasn't half?
- Did you find a way to make half that surprised you?
- How did you prove to yourself that the two parts were equal?

As a class, explore the other shapes that students partitioned, asking the following questions:

- What shapes did you try to cut in half? What ways did you find?
- Which shapes were easier to cut in half, and which were harder? Why?
- What ways did you find that surprised you?
- What observations did you make about the shapes and half?
- Did the halves you made with these shapes follow our definition of half? Why or why not? Do we need to revise our definition?

In this whole discussion, be sure to attend to mistakes and showcase them as constructive places for the class to learn more about what makes a half and the properties of shapes.

Extend

Repeat this activity using the same figures, but invite students instead to think about the idea of "fourth" or a "quarter." Ask, How can we cut the rectangle into fourths? What is a fourth? By exploring different ways to cut shapes into fourths, develop a definition for *fourths* that centers again on making equal parts. Students may begin with ways to cut the rectangle in half, by simply cutting it in half twice.

In your discussion, ask, How are fourths related to halves? Be sure that students can see that fourths are half of a half, or that subdividing into fourths is the same as cutting something in half and then half again. You can use the same shape sets in this extension, and students will benefit from seeing these shapes first cut in half and then in fourths to build an understanding of the relationship between half and fourth.

Students may think of fourths by drawing the diagonals of the rectangle. This presents an interesting discussion. If we cut along both diagonals, the result provides two pairs of triangles, one pair very different than the other. Students may want to explore the idea of equal by cutting again so that they can match the pieces. The series of images here shows a progression of cuts to justify that the triangles are equal in area. The process of physically showing that the areas are equal requires rotations and flips that are a great extension of this conversation.

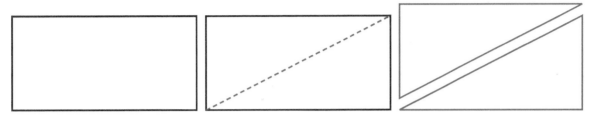

Students may choose to cut the rectangle in half by cutting along one diagonal. A series of flips or a rotation will show the two triangles have equal area.

With cuts along each diagonal, two pairs of triangles are created, the green and pink triangles do not look equal in area. If another set of cuts are made, the triangle marked "a" can be flipped or rotated to show that it is equal to triangle "b."

Look-Fors

- **Are students making two (or four) parts, or two (or four) equal parts?**
  Students' everyday understandings of half, in particular, can be imprecise. Students may think that half can be any way of cutting a figure into two parts, and this conception can carry over to fourths if not addressed. It may be useful to introduce the idea of "fair shares" if students are struggling to think about equal parts. You might say that when we cut something in half, we should be making two fair shares. You might ask students to imagine that the rectangle is a treat or a special piece of paper, and that they need to find ways to share it fairly with someone else. Ask, How could you cut this rectangle so that it could be shared fairly between two people? Use students' own ideas to build a definition of half as two equal parts.

- **Are students verifying with precision that parts are equal?** First graders are still developing fine motor control, so we have no expectation that folding, cutting with scissors, or drawing lines will be precise. However, we want students to be conceptually precise, meaning that they have a clear understanding that the portions they cut should be equal. Look for students checking for equality by putting shapes on top of one another or folding them together, and invite them to share these strategies in the discussion. Otherwise, ask students, How do you know these two (or four) parts are equal (or the same)? Notice also whether students use flipping and turning to assess whether the two halves or four fourths are the same. Students may need to be persistent to find an arrangement where their pieces align, and you may need to encourage them to keep trying if their initial efforts appear not to match.

- **Are students making connections between halves and fourths?** When students explore the extension to build an understanding of fourths, our aim is that they compare their work with halves. As you observe students working, you may see some cutting shapes into fourths by making four independent portions. These may be relatively imprecise because students are thinking about making four regions, rather than about how to decompose into fourths. Other students may be making half and then cutting those halves in half again. These types of strategies are much more likely to be precise because they are built on the meaning of fourth and its relationship to halves. Ask questions about how students knew that this type of approach would yield fourths. Be sure to highlight how students used the relationship between half and fourth to partition and why that relationship makes sense.

Reflect

What is a half?

How Could You Cut This Rectangle in Half?

Cutting Shapes 1

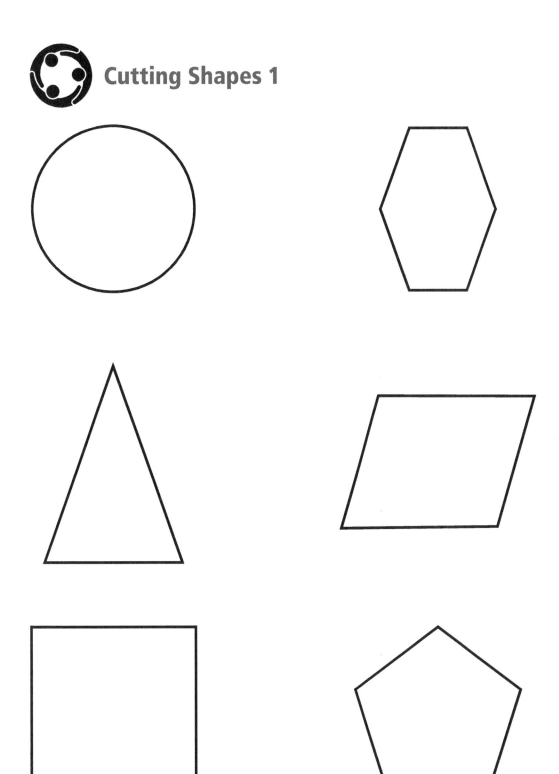

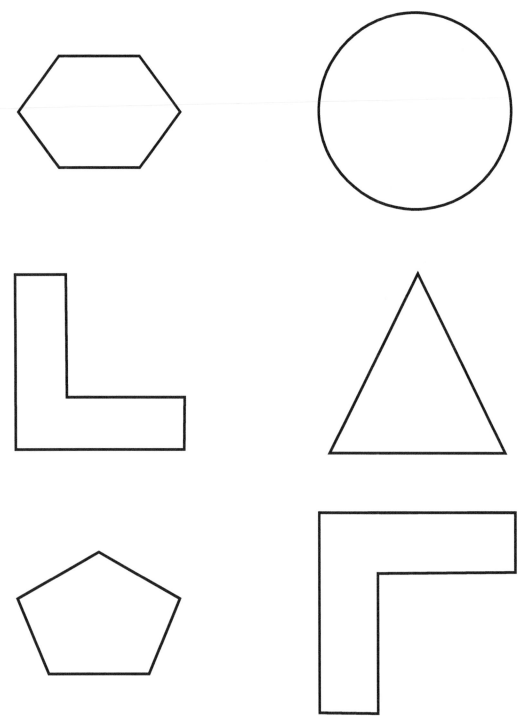

Analyzing Tangrams

Snapshot

Students play with halves and fourths by exploring the relationships between tangram pieces.

Connection to CCSS
1.G.3

Agenda

Activity	Time	Description/Prompt	Materials
Launch	5–10 min	Show the tangram pieces and ask, Can you find any pieces that are half of another piece? Invite a student to share an idea and how they know. Discuss both the half and the whole.	Tangrams, to display
Play	20–30 min	Partners play with tangram pieces to find as many different ways of representing half and fourth as possible, along with what they are half and fourth of—the whole. Partners trace or draw their findings as evidence to share.	Tangrams, per partnership
Discuss	15 min	Invite students to share a variety of examples of halves and wholes, fourths and wholes, and their evidence. Discuss the relationships they found between the pieces.	Tangrams, to display

To the Teacher

In this activity, we return to tangrams, which we used all the way back in Big Idea 1 to compose and decompose shapes. Here we use the relationships between the pieces to find multiple ways to represent half and fourth. What is critical, however, is attending to the whole. When students see half, what is it half of? When students see a fourth, what is it a fourth of? Fundamentally, fractions are a relationship between a part and a whole, and we cannot talk about the part without acknowledging that it is a part of a particular whole. In this activity, students get to define both the part and the whole as they look for half and fourth.

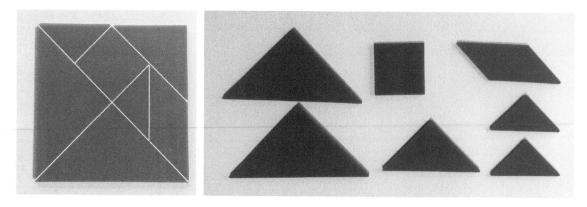

(Left) A tangram set makes a square. (Right)
A tangram set consists of seven pieces.

Students may notice that there is a relationship between the triangles in the tangram set. One relationship is that there are two of the large and two of the small triangles. If two identical triangles are joined, they make a whole where each triangle is half. Alternatively, students might see that the larger triangles are formed using two of the triangles that are next smaller in size; that is, the large triangle is composed of two medium triangles, and the medium triangle is composed of two small triangles.

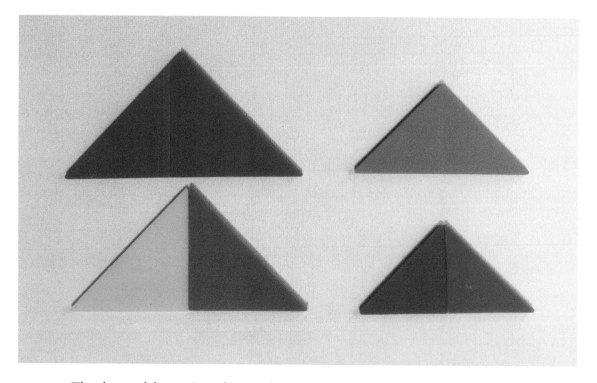

The large blue triangle can be made by two medium triangles.
The medium red triangle can be made by two small triangles.

However, students do not need to stick with defining the whole, or even the half or fourth, as a single piece. Students might notice that if they form all of the shapes into the tangram square, the two larger triangles are half of the square, and one of those triangles is a fourth of the square. Encourage students to think creatively about how to make half and fourth by playing with the parts and the whole. The number of solutions to this task is virtually unending, and if you offer students additional sets of tangrams to draw pieces from, then the possibilities are limitless.

Whether you have a large supply of tangram sets or not, you may want to engage your students in creating their own sets through paper folding. The act of folding paper to create shapes can help students see and think about the relationships between them. We included a set of folding directions in the Play activity in Big Idea 1 if you'd like to try this with your class as part of the launch.

Activity

Launch

Launch the activity by showing students a set of tangram pieces on the document camera. Ask, Can you find any pieces that are half of another piece? Give students a chance to turn and talk to a partner about where they might see half in these shapes. Invite a student to come up and show an example of half that they might see in the pieces. Be sure to draw attention to both the half and the whole being described. *Half* as an idea only makes sense when we know what it is *half of*.

Play

Provide partners with a set of tangrams and tools for recording. Partners work together to find as many ways as they can to show half and fourth with the tangram pieces, focusing on the relationships between the pieces. Students explore the following questions:

- Where can you find pieces that show *half*? Half of what?
- Where can you find pieces that show *fourth*? A fourth of what?
- How can you be sure that you have found half or fourth? What is your evidence?

For each half or fourth that they find, students record the shapes by tracing or drawing to show what pieces make half (or fourth) and what they are half (or fourth) of. Encourage creativity. If students compose shapes to make a whole or a half, celebrate these inventive ways of think about half.

Discuss

As a class, discuss the following questions:

- What were you looking for when you searched for half and fourth?
- What strategies did you use for finding half or fourth?
- What examples of half did you find with the tangram pieces?
- How did you know that these were half? Half of what?
- What examples of fourth did you find with the tangram pieces?
- How did you know that these were a fourth? A fourth of what?
- What relationships between the pieces did you find?

Be sure to focus on students' evidence and strategies for proving that they had equal parts, including stacking pieces on top of one another, folding paper, tracing, rotating, and flipping.

Look-Fors

- **Are students thinking flexibly about the whole?** If students are familiar with the tangrams as a square, they may think of this square as the only possible whole. Similarly, students may also see the whole as the collection of seven shapes. If the whole is fixed in these ways, then there are still multiple ways of finding half and fourth, but it limits the possibilities. If you notice that students are fixed on finding half (or fourth) of the full tangram square or shape set, ask questions to help open up the possibilities, such as, What else could be the whole? Alternatively, you might notice and ask, I see that you're looking for shapes that are half of the square. What else could you find half of?

- **Are students using halves to find fourths?** Part of this big idea is developing the relationship between half and fourth, as we began to do in the Visualize activity. As you circulate, look for students who are modifying their solutions for half to find fourth. For instance, if students notice that the medium triangle is half of the large, they could decompose the medium triangle into two small triangles to help find fourth. Be sure to have students who use these strategies share their thinking in the discussion. Alternatively, if students find many solutions for half, but struggle to find fourths, you might support using the relationship between them by asking, How could you use your solutions for half to help you find fourths?

- **Are students using multiple pieces to create halves, fourths, and wholes?**
  You may notice that it is natural for students to begin this task by looking for a single piece that is half or fourth of another single piece. However, many more possibilities open up if students begin to construct halves, fourths, and wholes out of multiple pieces. For instance, the medium triangle can be seen as a fourth of the two large triangles, if the two large triangles together form the whole. Alternatively, the medium and two small triangles together form half of the two large triangles.

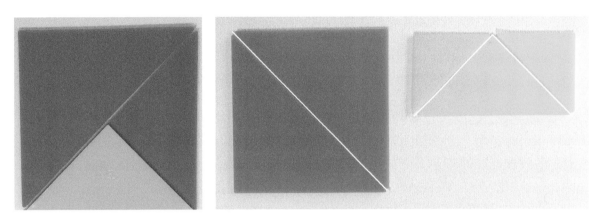

(Left) The medium yellow triangle is one fourth of the square made by two large triangles. (Right) One medium and two small triangles are one half the size of the square made by two large triangles.

To help students think about these ideas, you might focus on a whole they have selected and ask, What other shapes could be half (or a fourth) of this whole? You could also notice and ask, I see you're using a single shape to be your whole. What would happen if you made a whole out of more than one shape? What would the half (or fourth) be?

- **Are students seeing half when the two regions are not the same shape?**
  In the Visualize activity, we began to open up conversation about what half means, particularly by naming equal or same parts. However, students may have a conception of *equal* or *same* in this context as congruent, or identical, shapes. However, halves do not need to look identical; they need to have the same area. Although first graders have not yet had the opportunity to develop the concept of area, they can begin to think about half through area in this activity. For instance, in the image here, the square is half of the whole made of the square and two triangles. To provide evidence for this idea, students may

need to transform the shape, joining the two triangles to see that they form a shape that is identical to the square. However, even when they appear different from the square, they are still equal to it.

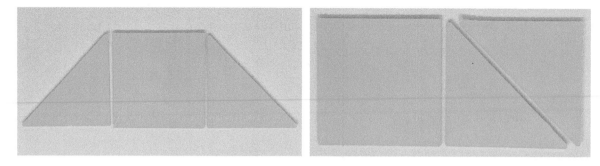

A square and two small triangles make a trapezoid that is the same size as a rectangle made of the same three shapes.

This represents a big conceptual leap in thinking about the meaning of equal parts. If you notice anyone trying this idea in their solutions, press them to provide evidence using the tangram pieces themselves. Then invite them to share their idea with the class and invite others to debate whether the parts have to look the same to be equal.

Reflect

Look around your classroom. Where can you see half (or fourth) right now? How do you know?

What's in a Flag?

Snapshot

Students investigate the use of equal parts in flags from around the world, exploring the different ways halves, thirds, and fourths can look and debating what makes parts equal.

Connection to CCSS
1.G.3

Agenda

Activity	Time	Description/Prompt	Materials
Launch	10 min	Remind students of their previous work with making equal parts, particularly with rectangles. Tell students that flags are rectangles we often partition to make designs. Show the image of the flag of Benin and ask, How did they partition this rectangle to make the flag of Benin? Invite students to turn and talk and share ideas about whether the flag shows equal parts. Name these parts as *thirds*.	Flag of Benin sheet, to display
Explore	30+ min	Partners investigate which flags are partitioned into equal parts, including halves, thirds, fourths, and any other equal partitioning. Students sort the flags into groups and then consider which flags are not partitioned into equal parts.	• World Flag Bank, per partnership, cut apart or as sheets • Make available: scissors, sticky notes • Optional: access to World Fact Book online or printed out

Activity	Time	Description/Prompt	Materials
Discuss	10–15 min	Create a display using students' ideas showing which flags are partitioned into 2, 3, 4, or more equal parts. Discuss students' evidence for these decisions, the strategies they used for determining equal parts, the relationships between different flags with the same number of equal parts, and other observations.	Display space with labels for flags with halves, thirds, and fourths
Extend	25+ min	Students design their own flags, for their family, class, school, or town, using equal parts. Students consider how to construct equal parts and what the parts of their flags will represent.	Make available: construction paper, colors, glue, tape, grid paper (see appendix), and rulers

To the Teacher

In this investigation, we invite students to explore equal parts with flags from around the world. In the previous two activities, we focused on developing ideas about halves and fourths, but with flags, three equal parts are by far the most common. Drawing on what students have learned about two and four equal parts, we invite students to build ideas about thirds. In the launch, students are presented with the flag of Benin, which is partitioned into three parts, but it is not obvious whether these parts are equal. This selection is intended to focus attention on how we know whether regions are equal and to highlight that parts can be equal even if they are a different shape or are not oriented in the same way. Encourage debate around this issue to support deeper reasoning about the idea of equal parts.

For the exploration, we have provided a bank of flags for students to explore and sort, but this collection is necessarily only a sampling of national flags. For a full set, you and your students can explore the World Fact Book website (https://www.cia.gov/library/publications/the-world-factbook/docs/flagsoftheworld.html), which is assembled and curated by the US government. The site includes world flags coupled with data about and maps of each country. While the mathematics of this investigation centers on the flags, encourage students to wonder about the places these flags represent. Students will naturally want to know where the countries are and how to pronounce their names, and they may also be curious about other things, such as, What languages do they speak? How far away is that from here? What does it look like there? The World Fact Book website (and the print version) can help with many of these questions and may inspire more.

One issue that will arise in examining flags is how to make sense of the many decorations that often adorn the field, such as emblems, crests, seals, and symbols. For the purposes of this activity, we encourage you to focus students' attention only on the field—the region in the background of the flag—and to disregard decorative features. For instance, when students are examining the flag of Eritrea, working with the red, blue, and green triangles is complex enough, and there is no mathematical need to complicate the question of equal parts with the gold wreath and olive branch.

Activity

Launch

Launch the activity by reminding students that you have been exploring how to cut shapes into equal parts, and you have focused on halves and fourths. In the Visualize activity, you spent a lot of time focusing on rectangles, in part because rectangles are used so much in our world. Tell students that one place we use rectangles is in flags. Show students the flag of Benin, a country in West Africa, and ask, How did they partition (or cut) this rectangle to make the flag of Benin? What do you notice? Give students a chance to turn and talk, and then discuss students' observations. Be sure to focus on whether students think the rectangle has been cut into equal parts. Ask, How could we prove whether the parts are equal? Provide students with a word to describe three equal parts: *thirds*.

Explore

Provide partners with access to images of flags either through the World Flag Bank sheets or using online resources such as the World Fact Book website (https://www.cia.gov/library/publications/the-world-factbook/docs/flagsoftheworld.html). If you provide students with the World Flag Bank sheets, you can choose to cut these into individual flags ahead of time or offer access to scissors for students who would like to do so themselves. Partners explore the flags, sorting them either into groups or by labeling, as they investigate the following questions:

- Which flags are made of equal parts?
- Where can you see halves? How do you know?

- Where can you see fourths? How do you know?
- Where can you see thirds? How do you know?
- What other kinds of equal parts do you see?
- What flags do not use equal parts? How do you know?

Encourage students to annotate each flag to show where they see the equal parts and how they would describe those parts with words. As you circulate, press students on their reasoning, particularly about what makes parts equal. If students sort the flags into physical groups, they may also want to label these groups using sticky notes or other tools.

Discuss

Create a display of flags that have halves, thirds, fourths, and any other equal parts students identified by asking students to contribute flags they found with 2, 3, 4, or more equal parts. Include students' labels for which regions of the flag have each of these portions. As a class, discuss the following questions:

- What did you notice when you looked at flags for equal parts?
- How did you know whether a flag had equal parts? What strategies did you use?
- What flags do not use equal parts? How do you know?
- How are flags with halves (or thirds or fourths) the same? How are they different?
- What patterns do you notice?
- What other ways of using equal parts did you find?

Encourage discussion of any flags about which there is disagreement regarding whether the flag is partitioned into equal parts or whether regions of the flag are equal. Be sure to ask the class questions to promote reasoning, such as, How could we prove whether these parts are equal or not?

When students discuss flags, encourage them to use the names of the countries to which they belong and create space for them to wonder about these places. The World Fact Book website can provide quick answers to questions such as, Where is this place? and How do you say this name?

Extend

Invite students to design a flag for their family, class, school, or town that uses equal parts. As they work to create a plan, students consider the following questions:

- What equal parts will your flag have?
- How will you design your flag?
- What do the parts of your flag represent? What is the meaning of the parts you've chosen?
- How will you make sure that your parts are equal?

Provide students with access to construction paper, colors, glue, tape, grid paper (see appendix), and rulers for their designs. Students can share their designs with the class and discuss the challenges they faced in partitioning their flags into equal parts.

Look-Fors

- **How are students determining equal parts?** Students may use a variety of strategies to determine which flags have equal parts. The key, however, is whether these strategies reflect equivalence or simply partitioning. For instance, it is not enough to determine that a flag has three parts, or that these parts look roughly equal. Students need evidence to confirm that the regions are the same size. They might fold or cut a flag into parts and then lay these parts on top of one another or hold them up to the light to see if they are indeed the same. Students might measure with their fingers or using a non-standard unit, such as a pencil eraser, to compare the width of stripes. Whatever the strategy, look for evidence that students are focusing on whether the regions are the same size.

- **How are students making sense of equal parts that are not oriented or shaped the same?** While students may try the strategies described in the previous Look-For to determine equivalence, these strategies may miss equal parts that do not immediately align. For instance, not all equal parts are oriented in the same way, as in the flag of Benin. Folding, in this case, may lead students to think that the parts are not equal. Students would need to either cut these regions out and rotate them or reason about the ways that the flag is partitioned to make claims about equivalence. In the case of the flag of Eritrea, students may rightly point out that the regions are not equal, yet the red triangle is half of the flag, and each of the green and blue regions is a fourth. Reasoning about these regions requires deeper thinking about what half and fourth mean. You might ask, Is there a way to name how much of the flag is red (or green or blue)? In these more complex cases, it is not critical that students come to a conclusion about the flag but rather that they create intriguing places for mathematical discussions and debates.

- **How are students interpreting the flag and its features?** Some flags, like the flags of Mauritius and Benin, have only a partitioned field with no decoration. This simplicity makes the question of equal parts clear. However, most flags are adorned in some way, and these features can be distracting. Students may struggle to know which features to ignore and which to consider. Students may also be drawing on their spatial relations to see flags as layered, instead of partitioned. For instance, with the flag of the Central African Republic, students may wonder whether they should ignore or include the star at the upper left. They may also see the horizontal stripes as fourths with a vertical red stripe layered on top. Both of these interpretations change the determination of whether the flag is partitioned equally. If students have surprising ideas about whether a flag is partitioned equally, ask questions to uncover how they are seeing and interpreting the flag. You may uncover that they are decomposing the flag in an entirely novel way. Make their interpretation part of their answer, as in, "If you consider the star, then the flag is not partitioned equally." This leaves room for other interpretations and answers.

- **Are students noticing structures that repeat across flags?** Many flags share a common structure, with only color or decoration distinguishing them. The most common structure is a field partitioned vertically into three equal parts. This is the case with the flags of Cameroon, Chad, France, Guinea, Ireland, Italy, Mali, Mexico, Nigeria, Peru, Romania, and Senegal, among many others. Three equal horizontal stripes is similarly common, including the flags of Azerbaijan, Bolivia, Bulgaria, Burma, Egypt, El Salvador, Gabon, Honduras, Lithuania, Malawi, Paraguay, Sierra Leone, and Yemen, among many others. First, it is intriguing to notice these patterns and wonder about why flags across the world share structures. Second, noticing this pattern can enable students to use their thinking from one flag to reason about another. For instance, once students determine that the flag of Peru has three equal parts, they can simply compare it to the flag of Mali to find that they are similar. If you see students starting with each flag anew, you might ask, Have you seen a flag like this before? Is there any way that could help you? You might also simply ask students what patterns they see among the flags. Patterns can be easier to discern the more flags students have seen, so be sure to inquire toward the end of the investigation and in the discussion.

Reflect

What other places in our world do you often see rectangles partitioned into equal parts? Sketch and label some examples.

Flag of Benin

Mindset Mathematics, Grade 1, copyright © 2021 by Jo Boaler, Jen Munson, Cathy Williams. Reproduced by permission of John Wiley & Sons, Inc. *Source:* 197734370 admin_design/Shutterstock.com.

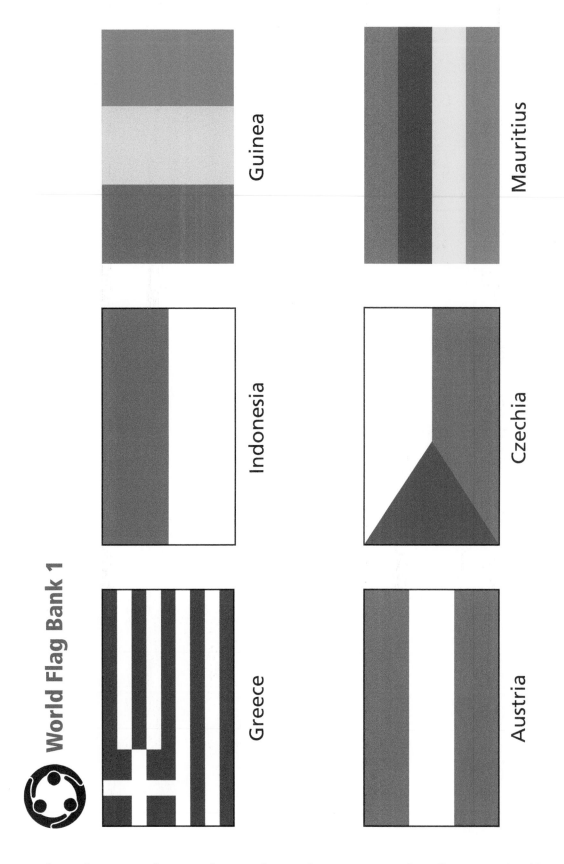

World Flag Bank 1

Greece

Indonesia

Guinea

Austria

Czechia

Mauritius

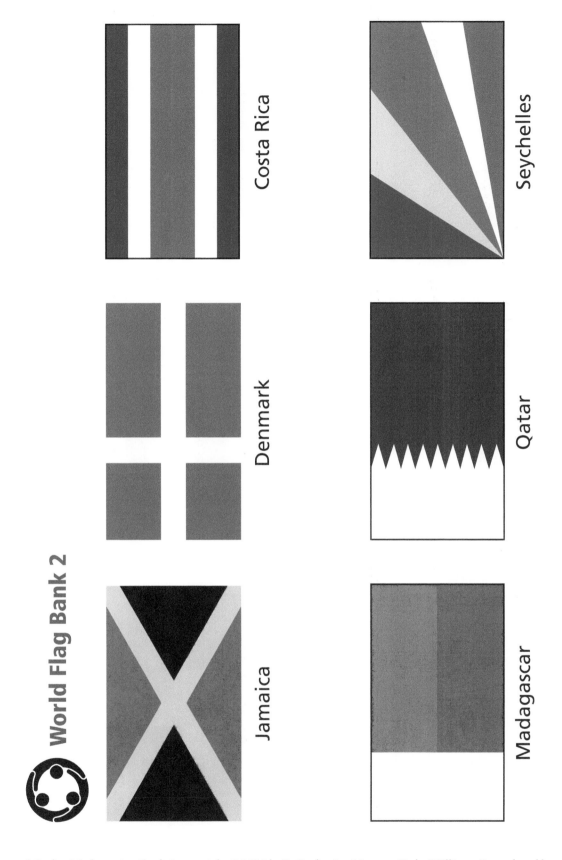

World Flag Bank 2

Costa Rica

Seychelles

Denmark

Qatar

Jamaica

Madagascar

Mindset Mathematics, Grade 1, copyright © 2021 by Jo Boaler, Jen Munson, Cathy Williams. Reproduced by permission of John Wiley & Sons, Inc. *Source:* 197734370 admin_design/Shutterstock.com.

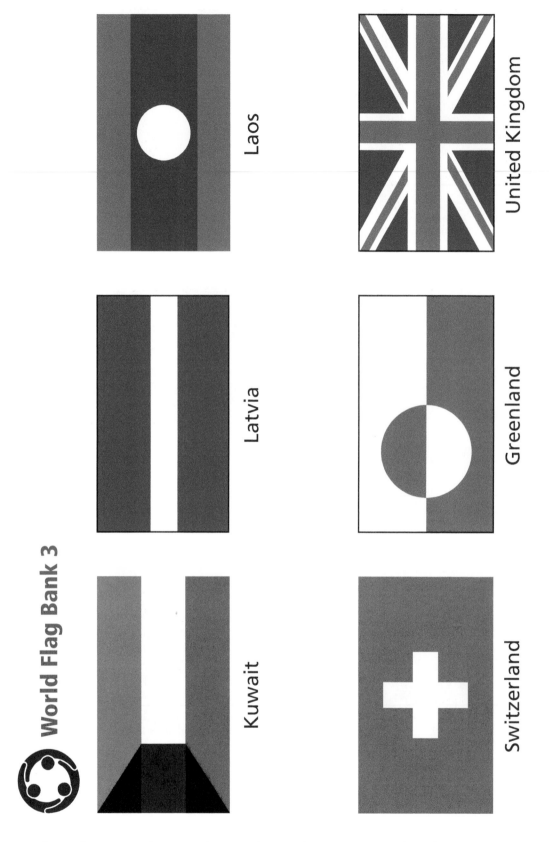

Laos

United Kingdom

Latvia

Greenland

Kuwait

Switzerland

Mindset Mathematics, Grade 1, copyright © 2021 by Jo Boaler, Jen Munson, Cathy Williams. Reproduced by permission of John Wiley & Sons, Inc. *Source:* 197734370 admin_design/Shutterstock.com.

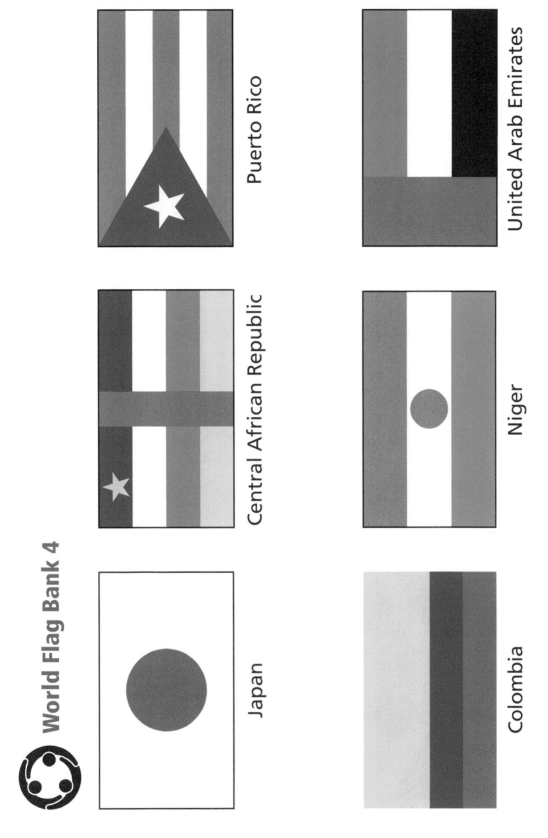

World Flag Bank 4

Puerto Rico

United Arab Emirates

Central African Republic

Niger

Japan

Colombia

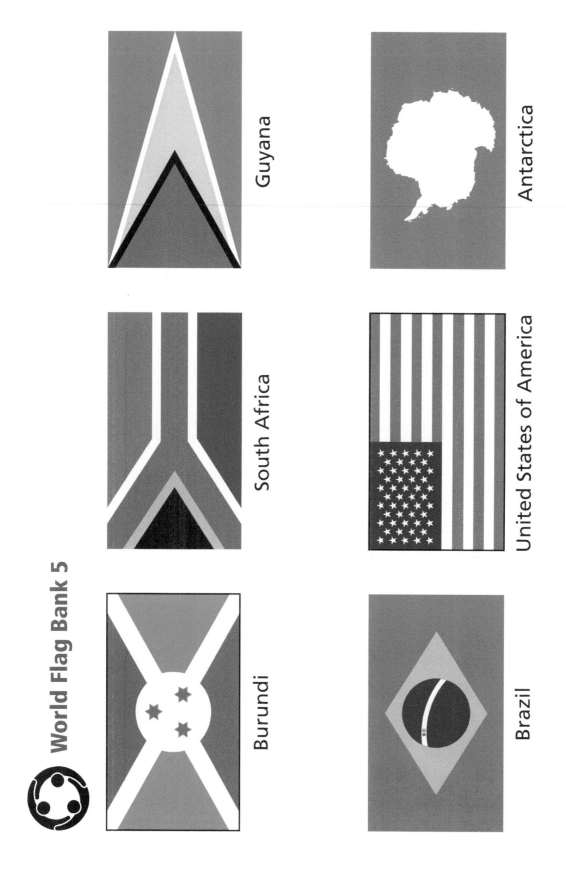

World Flag Bank 5

Guyana

Antarctica

South Africa

United States of America

Burundi

Brazil

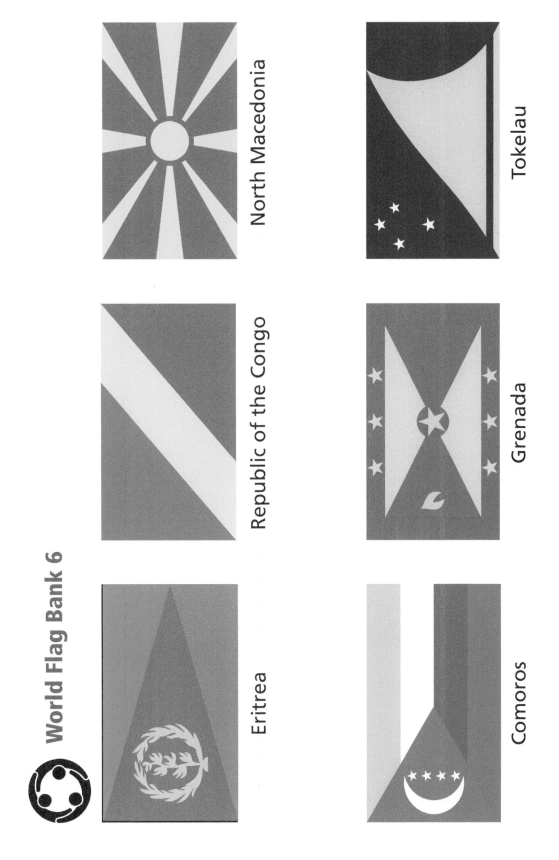

World Flag Bank 6

North Macedonia

Tokelau

Republic of the Congo

Grenada

Eritrea

Comoros

Appendix

1"Grid Paper

Appendix

Hundred Chart

1	2	3	4	5	6	7	8	9	10
11	12	13	14	15	16	17	18	19	20
21	22	23	24	25	26	27	28	29	30
31	32	33	34	35	36	37	38	39	40
41	42	43	44	45	46	47	48	49	50
51	52	53	54	55	56	57	58	59	60
61	62	63	64	65	66	67	68	69	70
71	72	73	74	75	76	77	78	79	80
81	82	83	84	85	86	87	88	89	90
91	92	93	94	95	96	97	98	99	100

Grid Paper

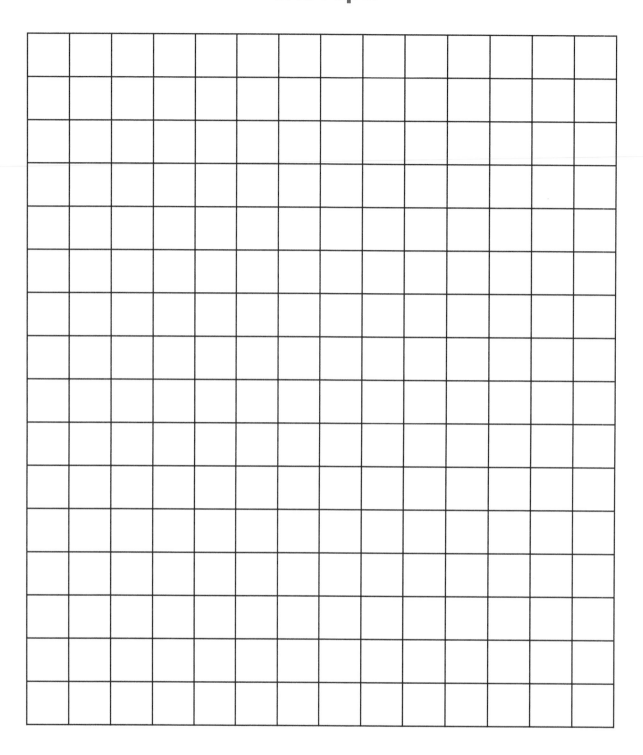

Appendix

About the Authors

Dr. Jo Boaler is a professor of mathematics education at Stanford University, and the cofounder of Youcubed. She is the author of the first MOOC on mathematics teaching and learning. Former roles have included being the Marie Curie Professor of Mathematics Education in England, a mathematics teacher in London comprehensive schools, and a lecturer and researcher at King's College, London. Her work has been published in the *Times,* the *Telegraph,* the *Wall Street Journal,* and many other news outlets. The BBC recently named Jo one of the eight educators "changing the face of education."

Jen Munson is an assistant professor of learning sciences at Northwestern University, a professional developer, and a former classroom teacher. She received her PhD from Stanford University. Her research focuses on how coaching can support teachers in growing their mathematics instructional practices and how teacher-student interactions influence equitable math learning. She is the author of *In the Moment: Conferring in the Elementary Math Classroom*, published by Heinemann.

Cathy Williams is the cofounder and director of Youcubed. She completed an applied mathematics major at University of California, San Diego before becoming a high school math teacher for 18 years in San Diego County. After teaching, she became a county office coordinator and then district mathematics director. As part of her leadership work, Cathy has designed professional development and curriculum. Her district work in the Vista Unified School District won a California Golden Bell for instruction in 2013 for the K–12 Innovation Cohort in mathematics. In Vista, Cathy worked with Jo changing the way mathematics was taught across the district.

Acknowledgments

We thank Jill Marsal, our book agent, and the team at Wiley for their efforts to make these books what we'd imagined. We are also very grateful to our Youcubed army of teachers. Thanks to Robin Anderson for drawing the network diagram on our cover. Finally, we thank our children—and dogs!—for putting up with our absences from family life as we worked to bring our vision of mathematical mindset tasks to life.

Index

Collections of small objects, 19

Color-coding, 6–7, 205, 207

Colors, 19, 219, 222–223

Combinations, 190, 196

Common Core curriculum standards, 9, 281

Communication, 120, 128–129

Comparison: modeling for, 234–235; of models, 85–86; observations from, 280; questions, 109, 112, 180, 208; of sizes, 57; of strategies, 186; tools, 166; understanding and, 111

Comparison questions, 109

Components, 38

Composing: decomposing and, 44, 186–187, 248, 251–257, 259–260, 281; routines for, 178; shapes, 32, 34; thinking about, 37

Computations, 81–82

Conceptual engagement, 4–5, 59, 78–79

Connections, 6–7

Consensus, 157, 270, 302

Convincing, 5–8, 23, 156

Cordero, Montse, 1

Counting: arrangements for, 178–179; buttons, 63; *Choral Counting and Counting Collections*, 62; consistency in, 180; data, 129; double-counting, 180–181; grouping and, 61, 181; during investigation, 232; observations and, 237–238, 256; Organizing and Counting a Collection, 61–67; with partners, 61–62, 186; patterns for, 73; representation of, 119; skip-counting, 213–219, 251–257; sorting and, 128; strategies for, 68; by students, 205; thinking with, 245; visualizations for, 59–60, 136

Creativity: creative thinking, 91–92; with displays, 269; encouraging, 225; for engagement, 91; for families, 15; with games, 189; inspiration from, 11–12; in journals, 5–6; in mathematics, 31, 259; with models, 99; with partners, 41, 137, 156, 159, 187, 220; planning and, 125–126; with schools, 34–35; sharing, 120; with solutions, 236–237; with storytelling, 105; strategies and, 44, 234; of students, 8, 300, 302–303; with tangram pieces, 260

Critical tasks, 3–4

Cubes, 267–275

Cuisenaire rods, 19, 220–227, 275–280

Curiosity, 255, 259

Cutting, 285

D

Danielson, Christopher, 23

Data: Displaying Data, 115–123; during exploration, 76–77; investigation of, 124–132; learning from, 130, 138; literacy, 107–108; for partners, 111–112; recording, 129; science, 110–111

Debates, 155, 284, 298

Decisions, 99

Decomposing: composing and, 44, 186–187, 248, 251–257, 259–260, 281; routines for, 178; shapes, 32, 34, 41; strategies for, 247; for students, 177, 181, 304; thinking about, 37

Deep questions, 15

Definitions: language for, 53; of questions, 81–82; with shapes, 285–286; understanding of, 56

Description: language for, 34–35, 38, 197–198; tools and, 195; of units, 279–280

Mathematics: books and, 103; in brain science, 3–4; challenges in, 156; childhood mathematical thinking, 62; Common Core curriculum standards in, 281; conceptual engagement with, 4–5, 59; creativity in, 31, 259; engagement with, 13–14; equal signs in, 133–135; *Fostering Algebraic Thinking*, 8; mathematical ideas, 227; mathematical language, 158, 216; mental arithmetic, 10–11; methods and, 9; questions about, 100; of shapes, 55; of space, 101; for students, 4–8, 81; thinking about, 103; *Thinking Mathematically*, 166; understanding of, 87, 231; visual, 17; "What Is Mathematical Beauty?," 9; wonder and, 300–301

Mathematics, The Science of Patterns (Devlin), 203

Mathematics for Human Flourishing (Su), 203

Meaning, 82, 268, 283

Measurement: Measuring Our World in Cubes, 267–274; Measuring with Cuisenaire

Rods, 275–280; Using Units to Measure Our World, 259–260; Which One Is Longer? for, 261–266

Memorization, 4–5

Menon, Venod, 175

Mental arithmetic, 10–11

Methods: in brain science, 5; explaining, 6–7; in How Many Do You See?, 23; mathematics and, 9; questions and, 14, 66

Missing Number Cards, 170–173

Missing values, 134–135, 140, 165–169, 208–209

Mistakes, 14

Misunderstandings, 133

Modeling: for comparison, 234–235; manipulatives for, 17, 94, 199; with partners, 165, 195; puzzles, 140–141; Representing and Modeling Joining and Separating Situations, 81–82; situations, 188–189; storytelling and, 94–95; by students, 247; tools for, 251, 253–254

Models: activities with, 104; comparisons of, 85–86; creativity with, 99; development of, 85; encouraging with,

235–236; for interpretation, 243; learning from, 168–169; with partners, 83, 155; physical, 87–88; questions about, 86; representation of, 86–87; snap cubes for, 102; understanding, 93, 95

Modifying, solutions, 256–257

N

Narration, 95

Neuroscience, 2–5, 9–14, 175

Noticing buildings sheet, 39

Nrich, 53

Numbers: activities with, 138; Balance Number Puzzles, 146–153; Balance Number sheets, 142–145; Building with Numbers within 20, 175–176; conceptual engagement of, 78–79; convincing and, 156; evidence and, 155; Finding Patterns in Numbers, 203–204; instructions for, 157; joining, 233; learning, 59–60; Missing Number Cards, 170–173; number lines, 10; *Number Talks*, 178; numerical patterns, 226–227; with partners, 177;

Patterns in the Hundred Chart, 205–212; patterns with, 72–73; in Skipping across the Hundred Chart, 213–219; strategies with, 189–190, 234–235; for students, 75, 189–190; values and, 102–104

O

Objects: activities with, 278; classroom, 59, 61, 63; engagement with, 125–126, 276; for group Work, 124; grouping, 114; observations of, 261; with partners, 115, 118, 267, 275, 278; questions about, 263; real-world, 109; sorting, 110–111; testing, 270–271

Observations: from books, 101; by class, 286; with colors, 222–223; from comparison, 280; counting and, 237–238, 256; about data, 129–130; with displays, 220; encouraging, 256; from evidence, 263; investigation of, 254–255; learning with, 203; of numerical patterns, 226–227; of objects, 261; of patterns, 275; questions, 207; recording, 278; of snap cubes, 267; strategies and, 77; by students, 198–199; visualizations and, 301

OECD. *See* Organisation for Economic Co-operation and Development

1" Grid Paper, 314

Online resources, 100

Open-ended problems, 244

Open-ended tasks, 246–247, 252

Organisation for Economic Co-operation and Development (OECD), 4

Organization: communication and, 120; of dice, 70, 72; Organizing and Counting a Collection, 61–67; Organizing the Natural World, 109–114; questions about, 254; sorting and, 125; by students, 116–117, 252–253; Tens and Ones Are Useful Ways to Organize, 59–60; of thinking, 104–105; tools, 19, 64; from visualizations, 65

Our Trash, Ourselves, 124–132

P

Paper, 87–88

Parents, 12

Park, Joonkoo, 175

Partial units, 266

Partitioning: in Half and Half-of-a-Half, 283–292; Partitioning Shapes into Equal Parts, 281–282; of rectangles, 299–304

Partners: activities with, 295; color-coding with, 205, 207; counting with, 61–62, 186; creativity with, 41, 137, 156, 159, 187, 220; data for, 111–112; decisions by, 99; dice for, 69, 188; discussion with, 206, 223–224; encouraging, 64, 207; evidence with, 293; exploration with, 215–216; extension for, 72–73; gallery walks with, 221, 225, 270; games for, 68, 157–158, 214, 216–217, 235–236; index cards for, 74; information and, 102; investigation in, 52, 55–56, 301–302; manipulatives for, 36, 197; measurement by, 261; modeling with, 165, 195; models with, 83, 155; numbers with, 177; objects with, 115, 118, 267, 275, 278; planning with,

263–264; predictions with, 269; problem-solving with, 243, 248; puzzles for, 136; questions for, 138–139, 179, 224–225; recording by, 71–72; repetition for, 157; sharing with, 139–140; solutions from, 53, 167; sticky notes for, 109; storytelling with, 89, 93; strategies with, 262; students as, 33; thinking with, 285

Patterns: action and, 130; Addition Table Patterns, 195–201; for counting, 73; Finding Patterns in Numbers, 203–204; in investigation, 176; *Mathematics, The Science of Patterns*, 203; with numbers, 72–73; numerical patterns, 226–227; observations of, 275; Patterns in the Hundred Chart, 205–212; predictions from, 78; questions about, 215; skip-counting and, 213; for students, 279; Weaving Patterns, 220–229

Paying attention, 23

Physical models, 87–88

PISA. *See* Program for International Student Assessment

Place value, 217; learning with, 231–232, 238; in Making a Dollar, 251–257; Playing with Place Value across Problem Types, 243–250; in Recess!, 233–242; Using Place Value to Add and Subtract, 231–232

Planning, 125–126, 263–264

Plastic gloves, 20

Plastic sheet protectors, 20

Play: activities for, 82, 116–117, 129, 277; in Analyzing Tangrams, 293, 295; dice for, 60; discovery and, 298; engagement from, 108; equations for, 134; games for, 176; for How Many Dots?, 68, 70–71; investigation in, 221; learning from, 204; in Measuring Our World in Cubes, 267, 269; open-ended tasks for, 252; Playing with Place Value across Problem Types, 243–250; Playing with Problem Types, 89–98; problem-solving in, 232; for Rolling Ten, 186, 188–189; in Skipping across the Hundred Chart, 213, 215–216; with spatial

puzzles, 281–282; for students, 13–14, 115; for True or False?, 155, 157–158

Pomo Basket image, 228

Pomo culture, 224–225

Precision, 205, 265–266, 271, 288

Predictions: debates and, 284; exploration of, 198; learning with, 197; with partners, 269; from patterns, 78; questions about, 283, 285; by students, 79, 263

Presentation, 42

Problem-solving: extension for, 244–245; joining and, 90–91; logic and, 103–104; open-ended problems, 244; with partners, 243, 248; in play, 232; separation and, 90–91; sharing of, 102; for solutions, 44–45, 104–105; strategies for, 84–85, 165; tasks, 237; thinking and, 235, 254–255; tools for, 252

Program for International Student Assessment (PISA), 4

Proof, 160

Puzzles: Balance Number Puzzles, 146–153; in extension, 206; historical, 32, 260; learning with, 207; missing values in, 208–209; modeling, 140–141; for partners, 136; questions about, 137; relational, 134–135; spatial, 281–282; Tangram Puzzles lesson, 40–45; Tangram Puzzles sheets, 46–51

Sizes, 38, 57

Skip-counting, 213–219, 251–257

Skipping across the Hundred Chart, 213–219

Small-groups, 119

Snap cubes, 18, 75; colors and, 219; index cards and, 76–77; learning with, 141; for models, 102; observations of, 267; questions about, 269; recording tools and, 99

Solutions: from computations, 81–82; creativity with, 236–237; debates about, 298; displays for, 139; equivalence for, 56–57; from partners, 53, 167; problem-solving for, 44–45, 104–105; sharing, 179, 243, 246; by students, 256–257; values and, 169

Sorted Recycling image, 132

Sorting, 110–111, 113, 125, 128

Space, 101, 140, 214–215, 281–282

Squares, 18, 52–57

Statistics, 107

Sticky notes, 6, 68, 109–112

Stipek, Deborah, 62

Storm sheet, 48

Storytelling, 85, 89–98, 100–101, 105. *See also* Books

Strategies: chart paper for, 128; comparison of, 186; for counting, 68; creativity and, 44, 234; for decomposing, 247; development of, 243–244; grouping, 69; with numbers, 189–190, 234–235; observations and, 77; with partners, 262; with place value, 236; for problem-solving, 84–85, 165; questions about, 158, 168, 253; reasoning, 303; relational, 159; sharing, 65; for students, 75, 90, 261; thinking, 187; for understanding, 95

Structures, 304

Students: action by, 124–125, 128–129; activities for, 9; advice for, 31; books for, 116; categories for, 113–114; challenges for, 2–3, 284–285; classroom objects for, 59, 61, 63; cognitive effort for, 66; collaboration for, 12; combinations for, 196; counting by, 205; creativity of, 8, 300, 302–303; data literacy for, 107–108; decomposing for, 177,

181, 304; development of, 84, 89, 124, 175–176, 186; displays for, 76; drawing for, 118; engagement of, 4, 231–232, 259, 295; experimentation by, 296; exploration by, 53, 136, 221, 245–246; fractions for, 297; gallery walks for, 36–37; group work for, 21, 213; grouping by, 66, 72; interpretation by, 94; investigation by, 14–15, 165, 195, 220, 299; knowledge of, 284; lessons for, 3–4; low-floor, high-ceiling tasks for, 2–3; mathematics for, 4–8, 81; modeling by, 247; narration by, 95; number lines for, 10; numbers for, 75, 189–190; observations by, 198–199; open-ended problems for, 244; open-ended tasks for, 246–247; organization by, 116–117, 252–253; as partners, 33; patterns for, 279; play for, 13–14, 115; precision by, 271; predictions by, 79, 263; presentation for, 42; questions for, 35, 52, 130, 263–264;